# ESSENTIAL ENGINEERING EQUATIONS

**Syed A. Nasar**
**Clayton R. Paul**

Professors of Electrical Engineering
University of Kentucky
Lexington, Kentucky 40506

**CRC Press**
**Boca Raton Ann Arbor Boston**

**Library of Congress Cataloging-in-Publication Data**

Catalog record is available from the Library of Congress.



Direct all inquiries to CRC Press, Inc., 2000 Corporate Blvd., N.W., Boca Raton, Florida, 33431.



International Standard Book Number 0-8493-4263-5

Printed in the United States

# PREFACE

Four types of equations are commonly encountered in various branches of engineering. These equations are: linear, simultaneous algebraic equations; ordinary differential equations; partial differential equations; and, difference equations. In this book, one chapter is devoted to each of these four classes of equations, whose solutions are generally obtained in exact forms (such as analytical solutions). In addition, numerical methods of solving most of the general equations of all the four categories are included in a separate chapter. Examples are drawn from all the major branches of engineering. Thus, problems are included to illustrate the following: electric circuit theory, linear systems, electromagnetic field theory, mechanics, bending of beams, buckling of columns, twisting of shafts, vibration, fluid flow, heat transfer, and mass transfer. For prerequisites, the reader is expected to have had the usual courses in college physics and calculus. Familiarity with elementary vector operations is needed in the study of partial differential equations.

# CONTENTS

CHAPTER

# 1

# LINEAR, ALGEBRAIC EQUATIONS AND MATRIX ALGEBRA

## 1.1 Linear, Algebraic Equations

Linear, algebraic equations arise frequently in engineering in modeling linear, physical systems. Moreover, numerical methods of solving ordinary differential equations (Chapter 2), difference equations (Chapter 4) and partial differential equations (Chapter 3) also often result in linear, algebraic equations. The simplest case is one equation in one unknown:

$$ax = b \tag{1}$$

Here $a$ and $b$ are assumed to be known and the object is to solve for the unknown variable $x$. The solution is obvious and is given by $x = b/a$. Any of the variables $a$, $x$, and $b$ need not be constants but may be functions of some independent variables. For example, we may have $2t^2 x(t) = 3cos 4t$. In this example $a = 2t^2$ and $b = 3cos 4t$.

The equation is said to be algebraic in that no derivatives or integrals of the unknown $x$ appear. For example, $a_1\frac{dx(t)}{dt}+a_2x(t)=b$ would not be algebraic due to the $\frac{dx(t)}{dt}$ term. An equation is linear in that $x$ is not raised to a power greater than one. An example of a non-linear equation is $a_1x^2+a_2x=b$.

A set of simultaneous $n$ linear, algebraic equations in the $n$ unknowns $x_1, x_2, ..., x_n$ is of the form

$$
\begin{aligned}
a_{11}x_1 + a_{12}x_2 + \ldots + a_{1n}x_n &= b_1 \\
a_{21}x_1 + a_{22}x_2 + \ldots + a_{2n}x_n &= b_2 \\
\vdots \qquad\qquad & \quad \vdots \\
a_{n1}x_1 + a_{n2}x_2 + \ldots + a_{nn}x_n &= b_n
\end{aligned} \tag{2}
$$

It is possible to consider $m$ equations in $n$ unknowns. However, we generally strive to obtain the same number of equations and unknowns in modeling a physical system and we will assume this to be the case. This chapter will be devoted to the solution to (2).

## 1.2 Solution of Linear Equations by Gauss Elimination

The most efficient method of solving (2) either by hand or with the aid of a digital computer is the *Gauss elimination* technique. The heart of this technique is to reduce (2) to an equivalent set of equations (having the same solutions $x_1, ..., x_n$ as (2)) which are in triangular form:

$$
\begin{aligned}
c_{11}x_1 + c_{12}x_2 + c_{13}x_3 + \ldots + c_{1n}x_n &= d_1 \\
c_{22}x_2 + c_{23}x_3 + \ldots + c_{2n}x_n &= d_2 \\
\vdots \qquad\qquad\qquad & \quad \vdots \\
c_{(n-1)(n-1)}x_{n-1} + c_{(n-1)n}x_n &= d_{n-1} \\
c_{nn}x_n &= d_n
\end{aligned} \tag{3}
$$

Once (3) is obtained, the solution is obtained by "back substitution"; that is, solve the last equation for $x_n$, substitute $x_n$ into the next-to-last equation and solve for $x_{n-1}$, and continuing to "move upward" or "backward" to the next equation to substitute the previously determined unknowns and solving for the remaining unknowns of that equation.

The equations in (2) can be placed in the equivalent, triangular form of (3) by using the *simple elementary row* operation of replacing an equation with the sum of that equation and one of the other equations which as been multiplied by a constant.
For example, the equations

$$\begin{aligned} a_{11}x_1 + a_{12}x_2 &= b_1 \\ a_{21}x_1 + a_{22}x_2 &= b_2 \end{aligned} \tag{4}$$

are equivalent to

$$\begin{aligned} a_{11}x_1 + a_{12}x_2 &= b_1 \\ (a_{22} - \frac{a_{21}}{a_{11}}a_{12})\, x_2 &= b_2 \end{aligned} \tag{5}$$

The first equation of (4) has been multiplied by $-\frac{a_{21}}{a_{11}}$ and added to the second equation. This produces a zero coefficient of $x_1$ in the second equation. Comparing (5) to (3) we identify $c_{11} = a_{11}, c_{12} = a_{12}$, $c_{22} = a_{22} - \frac{a_{21}}{a_{11}}\, a_{12}$.

*Example 1-1.* Solve the following set of equations by Gauss elimination:

$$\begin{aligned} 2x_1 + x_2 - x_3 &= 4 \\ 3x_1 + 2x_2 + x_3 &= 2 \\ x_1 + x_2 - 2x_3 &= 1 \end{aligned}$$

*Solution*
Multiply the first equation by –3/2 and add to the second equation and multiply the first equation by –½ and add to the third equation:

$$\begin{aligned} 2x_1 + x_2 - x_3 &= 4 \\ + 1/2\, x_2 + \frac{5}{2}\, x_3 &= -4 \\ + 1/2\, x_2 - \frac{3}{2}\, x_3 &= -1 \end{aligned}$$

Now multiply the second equation of the resulting set by –1 and add to the third equation:

$$2x_1 + x_2 - x_3 = 4$$

$$\tfrac{1}{2}\, x_2 + \frac{5}{2}\, x_3 = -4$$

$$-4\, x_3 = 3$$

These equations are in triangular form. Solving for $x_3$ from the third equation of the triangular set we obtain

$$x_3 = -\frac{3}{4}$$

Substituting $x_3 = -\frac{3}{4}$ into the second equation of the triangular set and solving for $x_2$ yields

$$x_2 = -\frac{17}{4}$$

Substituting $x_3 = -\frac{3}{4}$ and $x_2 = -\frac{17}{4}$ into the first equation of the triangular set yields

$$x_1 = \frac{15}{4}$$

## 1.3 Uniqueness of Solutions

The equation in (2) may have a unique solution, no solution or an infinite number of solutions. To illustrate why this occurs, consider the set of two equations in two unknowns in (4). Each of these equations may be thought of as the equation of a line in two-dimensional space whose axes are $x_1$ and $x_2$. If these two equations intersect at only one point a unique solution exists as shown in Fig. 1-1(a). If the two lines are parallel and do not intersect then no solution (intersection) exists as shown in Fig. 1-1(b) and if the two lines are coincident then an infinite number of solutions (intersections) exist as shown in Fig. 1-1(c).

A set of three equations can be thought of as describing three planes in three-dimensional space. Intersections of these planes give solutions of the equations; if the three planes intersect at only one point, a unique solution to the corresponding equations exists. These concepts can be applied to $n$ equations representing $n$ surfaces in $n$-dimensional space although a graphical portrayal is not possible.

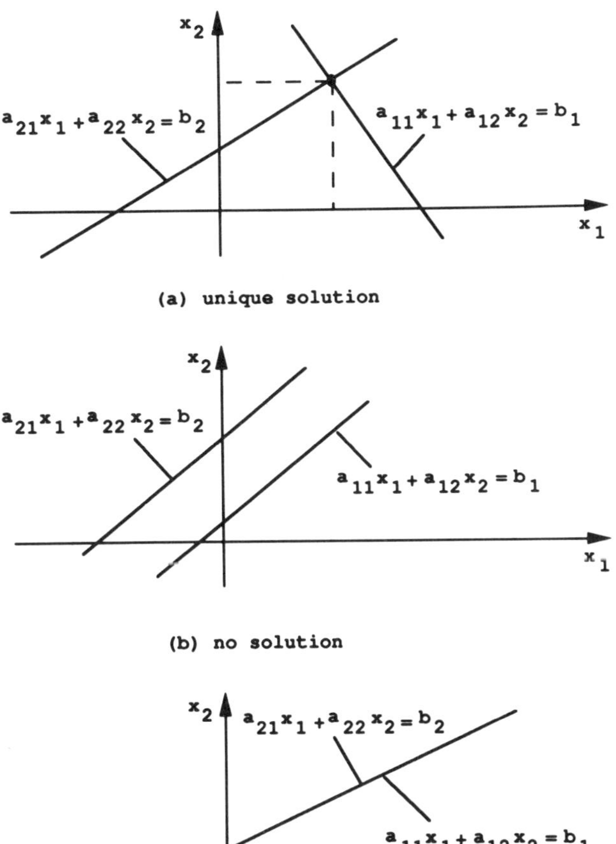


Figure 1.1

It is then easy to see that if all the $b_i$ elements on the right-hand side of (2) are zero, the $n$ surfaces represented by these $n$ equations all pass through the origin of the coordinate system with the result that the only possible solutions are either the unique, trivial solution with $x_1 = x_2 = \ldots = x_n = 0$ or an infinite number of solutions (two or more of the planes intersect along a line when n = 3).

An alternative method for solving (2) is Cramer's rule which is discussed in a subsequent section. However, Cramer's rule is considerably less efficient than Gauss elimination and is generally not employed for systems of more than three equations.

## 1.4 Matrix Algebra

A *matrix* is an ordered array of elements:

$$\mathbf{A} = \begin{bmatrix} a_{11} & a_{12} & \dots & a_{1n} \\ a_{21} & a_{22} & \dots & a_{2n} \\ \vdots & \vdots & & \vdots \\ a_{m1} & a_{m2} & \dots & a_{mn} \end{bmatrix} \tag{6}$$

A matrix with $m$ rows and $n$ columns is said to be $m \times n$ (pronounced "$m$ by $n$"). Each entry in the matrix is denoted as a lower case version of the matrix symbol $a_{ij}$ or, alternatively $[A]_{ij}$. Two subscripts attached to the element symbol denote the row and column location. If $m = n$, the matrix is said to be *square*. A matrix with only one row or only one column is given the name of a *vector*, and the location of an element will be denoted by only one subscript, i.e., $[V]_i = v_i$.

Two matrices are said to be equal if each of the corresponding elements of the two matrices are equal. Thus two matrices with different numbers of rows and/or different numbers of columns cannot be equal regardless of the entries.

Addition of two matrices may be defined if the two matrices are *conformable for addition*; that is, the two matrices have the same number of rows and the same number of columns. The sum of two $m \times n$ matrices $\mathbf{A}$ and $\mathbf{B}$ gives the $m \times n$ matrix $\mathbf{C}$ as the result,

$$\mathbf{C} = \mathbf{A} + \mathbf{B} \tag{7}$$

where

$$c_{ij} = a_{ij} + b_{ij} \tag{8}$$

and each entry in $\mathbf{C}$ is the sum of the corresponding entries in $\mathbf{A}$ and $\mathbf{B}$. For example, suppose $\mathbf{A}$ and $\mathbf{B}$ are $2 \times 3$, then

$$\mathbf{C} = \mathbf{A} + \mathbf{B}$$

$$= \begin{bmatrix} (a_{11} + b_{11}) & (a_{12} + b_{12}) & (a_{13} + b_{13}) \\ (a_{21} + b_{21}) & (a_{22} + b_{22}) & (a_{23} + b_{23}) \end{bmatrix} \tag{9}$$

Clearly it makes no difference in which order the two matrices are added, i.e., $\mathbf{A} + \mathbf{B} = \mathbf{B} + \mathbf{A}$.

The multiplication of a matrix by a scalar results in the multiplication of each entry in the matrix by the scalar.
For example,

$$\mathbf{C} = k\mathbf{A} \tag{10}$$

gives

$$c_{ij} = ka_{ij} \tag{11}$$

From this result we may define the subtraction of two matrices by multiplying the matrix to be subtracted by $-1$ and adding the two resulting matrices. For example,

$$\begin{aligned} \mathbf{C} &= \mathbf{A} - \mathbf{B} \\ &= \mathbf{A} + (-1)\,\mathbf{B} \end{aligned} \tag{12}$$

where

$$c_{ij} = a_{ij} - b_{ij} \tag{13}$$

The multiplication of two matrices as

$$\mathbf{C} = \mathbf{A}\,\mathbf{B} \tag{14}$$

gives the entry in $\mathbf{C}$ as

$$\begin{aligned} c_{ij} &= a_{i1}b_{1j} + a_{i2}b_{2j} + \dots + a_{in}\,b_{nj} \\ &= \sum_{k=1}^{n} a_{ik}\,b_{kj} \end{aligned} \tag{15}$$

Pictorially this rule becomes

$$\underset{\mathbf{C}}{\overset{\text{j-th column}}{i\text{-}th\ row \begin{bmatrix} & & \\ & c_{ij} & \\ & & \end{bmatrix}}} = \underset{\mathbf{A}}{i\text{-}th\ row \begin{bmatrix} a_{i1} & a_{i2} & \dots & a_{in} \\ & & & \end{bmatrix}} \underset{\mathbf{B}}{\overset{\text{j-th column}}{\begin{bmatrix} b_{1j} \\ b_{2j} \\ \\ b_{nj} \end{bmatrix}}} \tag{16}$$

and the $ij$ element of $\mathbf{C}$, $c_{ij}$, is the "dot product" of the $i$-th row vector of $\mathbf{A}$ and the $j$-th column vector of $\mathbf{B}$. Clearly the number of columns of **A** must equal the number of rows of **B** in which case **A** and **B** are *conformable for multiplication* in the order **A B**. Matrix **B** is *premultiplied* by **A** or, equivalently, matrix **A**, is *postmultiplied* by **B**. If **A** is $m \times n$ then **B** must be $n \times q$ and the result, $\mathbf{C} = \mathbf{A}\,\mathbf{B}$, is $m \times q$. Note that the product **B A** is not permissible if **A** is $m \times n$ and **B** is $n \times q$ unless $m = q$.

Note that matrix multiplication is, in general, not commutative; that is

$$\mathbf{A}\,\mathbf{B} \neq \mathbf{B}\,\mathbf{A} \tag{17}$$

Even though **A** and **B** may be square (and therefore conformable for multiplication as **A B** or **B A**) it is not necessarily true that $\mathbf{A}\,\mathbf{B} = \mathbf{B}\,\mathbf{A}$. If the product of two matrices yields the zero matrix, it is not necessarily true that either of the matrices is a zero matrix. Even though matrix multiplication is not commutative it is associative:

$$(\mathbf{A}\,\mathbf{B})\,\mathbf{C} = \mathbf{A}\,(\mathbf{B}\,\mathbf{C}) \tag{18}$$

Matrix addition is both commutative and associative:

$$\mathbf{A} + \mathbf{B} = \mathbf{B} + \mathbf{A}$$

$$\mathbf{A} + (\mathbf{B} + \mathbf{C}) = (\mathbf{A} + \mathbf{B}) + \mathbf{C} \tag{19}$$

Note that if $\mathbf{A} + \mathbf{B} = \mathbf{A} + \mathbf{C}$ then $\mathbf{B} = \mathbf{C}$. (Add $-\mathbf{A}$ to both sides.) Matrix multiplication is distributive with respect to addition

$$\mathbf{A}\,(\mathbf{B} + \mathbf{C}) = \mathbf{A}\,\mathbf{B} + \mathbf{A}\,\mathbf{C} \tag{20}$$

There are various special types of matrices. The $m \times n$ zero matrix with zeros in every position is denoted as ${}_m\mathbf{0}_n$. The square, $n \times n$ matrix, $\mathbf{1}_n$, is defined by

$$[\mathbf{1}_n]_{ii} = 1 \tag{21a}$$

$$[\mathbf{a}_n]_{ij} = 0 \qquad i \neq j \tag{21b}$$

The elements in the $i$-th row and $i$-th column of a square matrix are called the *main diagonal* terms. Thus the identity matrix has ones on the main diagonal and zeros elsewhere. Note that

$$\mathbf{1}_n\ \mathbf{A} = \mathbf{A} \tag{22}$$

if **A** is $n \times m$ and

$$\mathbf{A}\,\mathbf{1}_m = \mathbf{A} \tag{23}$$

The matrix A obtained by interchanging the rows and columns of A is referred to the transpose of A:

$$[\mathbf{A}^t]_{ij} = [\mathbf{A}]_{ji} \tag{24}$$

The following properties are easy to prove:

$$(\mathbf{A}^t)^t = \mathbf{A} \tag{25}$$

$$(\mathbf{A} + \mathbf{B})^t = \mathbf{A}^t + \mathbf{B}^t \tag{26}$$

$$(\mathbf{A}\,\mathbf{B})^t = \mathbf{B}^t\,\mathbf{A}^t \tag{27}$$

Note that the transpose of the product of two matrices is the product of the transpose of each matrix in reverse order.

*Example 1-2*

For the matrices

$$\mathbf{A} = \begin{bmatrix} 1 & 1 & 2 \\ 2 & 3 & 1 \end{bmatrix} \quad \mathbf{B} = \begin{bmatrix} 2 & 1 & 0 \\ 3 & 1 & 1 \end{bmatrix}$$

determine

$$\mathbf{A} + \mathbf{B}$$

$$\mathbf{A} - \mathbf{B}$$

$$(\mathbf{A} + \mathbf{B})^t$$

$$\mathbf{A}\,\mathbf{B}^t$$

*Solution*

$$\mathbf{A} + \mathbf{B} = \begin{bmatrix} 3 & 2 & 2 \\ 5 & 4 & 2 \end{bmatrix}$$

$$\mathbf{A} - \mathbf{B} = \begin{bmatrix} -1 & 0 & 2 \\ -1 & 2 & 0 \end{bmatrix}$$

$$(\mathbf{A} + \mathbf{B})^t = \begin{bmatrix} 3 & 5 \\ 2 & 4 \\ 2 & 2 \end{bmatrix}$$

$$= \mathbf{A}^t + \mathbf{B}^t$$

$$\mathbf{A\,B}^t = \begin{bmatrix} 1 & 1 & 2 \\ 2 & 3 & 1 \end{bmatrix} \begin{bmatrix} 2 & 3 \\ 1 & 1 \\ 0 & 1 \end{bmatrix}$$

$$= \begin{bmatrix} 3 & 6 \\ 7 & 10 \end{bmatrix}$$

$$= (\mathbf{B\,A}^t)^t$$

## 1.5 Determinants

The determinant of a square, $n \times n$ matrix **A** is denoted by $|\mathbf{A}|$ and given by

$$|\mathbf{A}| = \sum (\pm)\, a_{1i}\, a_{2j} \,...\, a_{nk} \tag{28}$$

where the sum is taken over all permutations of the second subscripts. A term is assigned a plus if $(i, j, ..., k)$ is an even permutation of $(1,2,...,n)$ and a minus sign if it is an odd permutation. Any rearrangement of the natural order $(1,2,...,n)$ of a set of numbers is regarded as a permutation. For example, (1,3,2,4) requires only one interchange (2 and 3) to achieve the natural order. On the other hand, (1,3,4,2) requires two permutations: (1,2,4,3) and (1,2,3,4) to achieve the natural order.

For example if **A** is $2 \times 2$ then

$$|\mathbf{A}| = a_{11}a_{22} - a_{12}\, a_{21} \tag{29}$$

and if **A** is $3 \times 3$,

$$|\mathbf{A}| = a_{11}a_{22}a_{33} - a_{11}a_{23}a_{32} - a_{12}a_{21}a_{33} + a_{12}a_{23}a_{31} + a_{13}a_{21}a_{32} - a_{13}a_{22}a_{31} \tag{30}$$

A more convenient method for evaluating a determinant is with the *Laplace expansion rule*. An expansion can be made along the $i$*-th* row of the determinant resulting in

$$|\mathbf{A}| = \sum_{k=1}^{n} a_{ik} (-1)^{i+k}\, M_{ik} \tag{31}$$

or along the $j$-th column resulting in

$$|A| = \sum_{k=1}^{n} a_{kj}\ (-1)^{k+j}\ M_{kj} \tag{32}$$

where $M_{ij}$ is the $(n-1) \times (n-1)$ determinant resulting after row $i$ and column $j$ in which $a_{ij}$ are located have been removed. $M_{ij}$ is said to be the *minor* of $a_{ij}$. The $M_{ij}$ factor with appropriate sign $(-1)^{i+j}$ is referred to as the $ij$ cofactor and denoted as

$$A_{ij} = (-1)^{i+j} M_{ij} \tag{33}$$

The proper sign of each cofactor, $(-1)^{i+j}$, can be found from the "checkerboard sign pattern":

$$\begin{matrix} + & - & + & - & + & - & \cdots \\ - & + & - & + & - & + & \cdots \\ + & - & + & - & + & - & \cdots \\ \bullet & \bullet & \bullet & \bullet & \bullet & \bullet & \\ \bullet & \bullet & \bullet & \bullet & \bullet & \bullet & \\ \bullet & \bullet & \bullet & \bullet & \bullet & \bullet & \end{matrix}$$

For example, for a $2 \times 2$ determinant we obtain

$$|A| = \begin{vmatrix} a_{11} & a_{12} \\ a_{21} & a_{22} \end{vmatrix} \tag{34}$$

$$= a_{11}\,|a_{22}| - a_{12}\,|a_{21}|$$

$$= a_{11}a_{22} - a_{12}a_{21}$$

by expanding along the first row. Expansion along the second column yields

$$|A| = -a_{12}\,|a_{21}| + a_{22}\,|a_{11}| \tag{35}$$

$$= -a_{12}a_{21} + a_{22}a_{11}$$

which is equivalent to (34).

For a $3 \times 3$ determinant we obtain, by expanding along the second row,

$$|\mathbf{A}| = \begin{vmatrix} a_{11} & a_{12} & a_{13} \\ a_{21} & a_{22} & a_{23} \\ a_{31} & a_{32} & a_{33} \end{vmatrix}$$

$$= -a_{21}\begin{vmatrix} a_{12} & a_{13} \\ a_{32} & a_{33} \end{vmatrix} + a_{22}\begin{vmatrix} a_{11} & a_{13} \\ a_{31} & a_{33} \end{vmatrix} - a_{23}\begin{vmatrix} a_{11} & a_{12} \\ a_{31} & a_{32} \end{vmatrix} \tag{36}$$

Expanding each $2 \times 2$ determinant in (36) we obtain the result in (29).

A square matrix **A** is said to be singular if $|\mathbf{A}| = 0$.

*Example 1-3*

For the $3 \times 3$ matrix **A**

$$\mathbf{A} = \begin{bmatrix} 1 & 2 & 3 \\ 4 & 5 & 6 \\ 7 & 8 & 9 \end{bmatrix}$$

determine $|\mathbf{A}|$.

*Solution*

Expanding along the second column

$$|\mathbf{A}| = -2\begin{vmatrix} 4 & 6 \\ 7 & 9 \end{vmatrix} + 5\begin{vmatrix} 1 & 3 \\ 7 & 9 \end{vmatrix} - 8\begin{vmatrix} 1 & 3 \\ 4 & 6 \end{vmatrix}$$

$$= -2\,(36 - 42) + 5\,(9 - 21) - 8\,(6 - 12)$$

$$= 0$$

Similarly, expanding along the third row

$$|\mathbf{A}| = 7\begin{vmatrix} 2 & 3 \\ 5 & 6 \end{vmatrix} - 8\begin{vmatrix} 1 & 3 \\ 4 & 6 \end{vmatrix} + 9\begin{vmatrix} 1 & 2 \\ 4 & 5 \end{vmatrix}$$

$$= 7\,(12 - 15) - 8\,(6 - 12) + 9\,(5 - 8)$$

$$= 0$$

## 1.6 Solution of Linear Equations by Cramer's Rule

Although the method of Gauss elimination is the most efficient technique for the solution of simultaneous, linear, algebraic equations, Cramer's rule is an alternate method which is suitable for hand calculations. To illustrate Cramer's rule, consider two equations in two unknowns:

$$a_{11}x_1 + a_{12}x_2 = b_1$$

$$a_{21}x_1 + a_{22}x_2 = b_2 \tag{37}$$

Solution by Gauss elimination will yield

$$x_1 = \frac{b_1 a_{22} - b_2 a_{12}}{a_{11}a_{22} - a_{12}a_{21}} \tag{38}$$

and

$$x_2 = \frac{-b_1 a_{21} + b_2 a_{11}}{a_{11}a_{22} - a_{12}a_{21}} \tag{39}$$

From this result we see a familiar pattern. If we write (37) in matrix form as

$$\underset{\mathbf{A}}{\begin{bmatrix} a_{11} & a_{12} \\ a_{21} & a_{22} \end{bmatrix}} \underset{\mathbf{X}}{\begin{bmatrix} x_1 \\ x_2 \end{bmatrix}} = \underset{\mathbf{B}}{\begin{bmatrix} b_1 \\ b_2 \end{bmatrix}} \tag{40}$$

we see that the denominators of the solutions for $x_1$ and $x_2$ are the determinant of the coefficient matrix, **A**. In addition we see that

$$x_1 = \frac{\begin{vmatrix} b_1 & a_{12} \\ b_2 & a_{22} \end{vmatrix}}{\begin{vmatrix} a_{11} & a_{12} \\ a_{21} & a_{22} \end{vmatrix}} \tag{41}$$

and

$$x_2 = \frac{\begin{vmatrix} a_{11} & b_1 \\ a_{21} & b_2 \end{vmatrix}}{\begin{vmatrix} a_{11} & a_{12} \\ a_{21} & a_{22} \end{vmatrix}} \tag{42}$$

Thus to find the solution for $x_i$ we take the ratio of the determinant of the coefficient matrix, **A**, with the $i$-th column of $|\mathbf{A}|$ replaced by the right-hand side vector **B** to the determinant of **A**.

This result holds for $n$ equations in $n$ unknowns and is known as *Cramer's rule*. For $n$ equations in $n$ unknowns we obtain

$$x_i = \frac{\begin{vmatrix} a_{11} & a_{12} & \dots & a_{1(i-1)} & b_1 & a_{1(i+1)} & \dots & a_{1n} \\ a_{21} & a_{22} & \dots & a_{2(i-1)} & b_2 & a_{2(i+1)} & \dots & a_{2n} \\ \vdots & & & \vdots & \vdots & \vdots & & \vdots \\ a_{n1} & a_{n2} & \dots & a_{n(i-1)} & b_n & a_{n(i+1)} & \dots & a_{nn} \end{vmatrix}}{\begin{vmatrix} a_{11} & a_{12} & \dots & a_{1(i-1)} & a_{1i} & a_{1(i+1)} & \dots & a_{1n} \\ a_{21} & a_{22} & \dots & a_{2(i-1)} & a_{2i} & a_{2(i+1)} & \dots & a_{2n} \\ \vdots & \vdots & & \vdots & \vdots & \vdots & & \vdots \\ a_{n1} & a_{n2} & \dots & a_{n(i-1)} & a_{ni} & a_{n(i+1)} & \dots & a_{nn} \end{vmatrix}} \tag{43}$$

or

$$x_1 = \frac{b_1 A_{1i} + b_2 A_{2i} + \dots + b_n A_{ni}}{|A|} \tag{44}$$

From this result it is easy to see the criterion for the existence of a unique solution to (2); $|A|$ must be nonzero. If the right-hand side elements of (2) are zero, $b_1 = b_2 = \dots = b_n = 0$, the set of equations is said to be *homogeneous*. The only unique solution to a homogeneous set of equations is the trivial solution, $x_1 = x_2 = \dots = x_n = 0$ (all surfaces intersect at the origin). From Cramer's rule we see that the unique (zero) solution to a homogeneous set of equations only if $|A| \neq 0$.

*Example 1-4*

For the set of three equations in Example 1-1 determine $x_1$, $x_2$ and $x_3$.

*Solution*

Writing the equations in matrix form we have

$$\underset{\mathbf{A}}{\begin{bmatrix} 2 & 1 & -1 \\ 3 & 2 & 1 \\ 1 & 1 & -2 \end{bmatrix}} \underset{\mathbf{X}}{\begin{bmatrix} x_1 \\ x_2 \\ x_3 \end{bmatrix}} = \underset{\mathbf{B}}{\begin{bmatrix} 4 \\ 2 \\ 1 \end{bmatrix}}$$

Expanding $|A|$ along the third row,

$$|A| = 1\begin{vmatrix} 1 & -1 \\ 2 & 1 \end{vmatrix} - 1\begin{vmatrix} 2 & -1 \\ 3 & 1 \end{vmatrix} - 2\begin{vmatrix} 2 & 1 \\ 3 & 2 \end{vmatrix}$$

$$= 1\,(1 + 2) - 1\,(2 + 3) - 2\,(4 - 3)$$

$$= -4$$

and

$$x_1 = \frac{\begin{vmatrix} 4 & 1 & -1 \\ 2 & 2 & 1 \\ 1 & 1 & -2 \end{vmatrix}}{|\mathbf{A}|}$$

$$= \frac{4\begin{vmatrix} 2 & 1 \\ 1 & -2 \end{vmatrix} -1\begin{vmatrix} 2 & 1 \\ 1 & -2 \end{vmatrix} -1\begin{vmatrix} 2 & 2 \\ 1 & 1 \end{vmatrix}}{-4}$$

$$= \frac{4\,(-4-1)\ -1\ (-4-1)\ -1\,(2-2)}{-4} = \frac{15}{4}$$

$$x_2 = \frac{\begin{vmatrix} 2 & 4 & -1 \\ 3 & 2 & 1 \\ 1 & 1 & -2 \end{vmatrix}}{|\mathbf{A}|}$$

$$= \frac{2\begin{vmatrix} 2 & 1 \\ 1 & -2 \end{vmatrix} -4\begin{vmatrix} 3 & 1 \\ 1 & -2 \end{vmatrix} -1\begin{vmatrix} 3 & 2 \\ 1 & 1 \end{vmatrix}}{-4}$$

$$= \frac{2\ (-4-1)\ -4\ (-6-1)\ -1\ (3-2)}{-4}$$

$$= \frac{-17}{4}$$

$$x_3 = \frac{\begin{vmatrix} 2 & 1 & 4 \\ 3 & 2 & 2 \\ 1 & 1 & 1 \end{vmatrix}}{|\mathbf{A}|}$$

$$= \frac{2\begin{vmatrix} 2 & 2 \\ 1 & 1 \end{vmatrix} -1\begin{vmatrix} 3 & 2 \\ 1 & 1 \end{vmatrix} +4\begin{vmatrix} 3 & 2 \\ 1 & 1 \end{vmatrix}}{-4}$$

$$= \frac{2\ (2-2)\ -1\ (3-2)\ +4\ (3-2)}{-4}$$

$$= -\frac{3}{4}$$

## 1.7 Matrix Inversion

The remaining matrix algebra operation to be defined is the inverse of a matrix. For a scalar a we define its inverse, $a^{-1}$, such that

$$a\ a^{-1} = a^{-1}a = 1 \tag{45}$$

This suggests the definition of the inverse of an $n \times n$ matrix $\mathbf{A}$, denoted as $\mathbf{A}^{-1}$, as the $n \times n$ matrix which has the property such that

$$\mathbf{A}\ \mathbf{A}^{-1} = \mathbf{A}^{-1}\ \mathbf{A} = \mathbf{1}_n \tag{46}$$

Denoting the entries in $\mathbf{A}^{-1}$ as $[A^{-1}]_{ij} = b_{ij}$ we need to solve for the $b_{ij}$'s which satisfy $\mathbf{AA}^{-1} = \mathbf{1}_n$; or

$$\underset{\mathbf{A}}{\begin{bmatrix} a_{11} & a_{12} & \dots & a_{1n} \\ a_{21} & a_{22} & \dots & a_{2n} \\ \vdots & \vdots & & \vdots \\ a_{n1} & a_{n2} & \dots & a_{nn} \end{bmatrix}} \underset{\mathbf{A}^{-1}}{\begin{bmatrix} b_{11} & b_{12} & \dots & b_{1n} \\ b_{21} & b_{22} & \dots & b_{2n} \\ \vdots & \vdots & & \vdots \\ b_{n1} & b_{n2} & \dots & b_{nn} \end{bmatrix}} = \underset{\mathbf{1}_n}{\begin{bmatrix} 1 & 0 & 0 & \dots & 0 \\ 0 & 1 & 0 & \dots & 0 \\ \vdots & & & \dots & \vdots \\ 0 & \dots & \dots & 0 & 1 \end{bmatrix}} \tag{47}$$

Treating the product of $\mathbf{A}$ and each column of $\mathbf{A}^{-1}$ and the associated column of $\mathbf{1}_n$ as a set of $n$ equations in $n$ unknowns:

$$\mathbf{A} \begin{bmatrix} b_{1i} \\ b_{2i} \\ \cdot \\ b_{ii} \\ \cdot \\ b_{ni} \end{bmatrix} = \begin{bmatrix} 0 \\ 0 \\ \cdot \\ 1 \\ \cdot \\ 0 \end{bmatrix} \tag{48}$$

and solving by Cramer's rule for $b_{1i}, b_{2i}, \dots, b_{ni}$ we obtain

$$b_{ki} = \frac{A_{ik}}{|A|} \tag{49}$$

Therefore

$$\mathbf{A}^{-1} = \frac{1}{|\mathbf{A}|} \begin{bmatrix} A_{11} & A_{21} & A_{31} & \dots & A_{n1} \\ A_{12} & A_{22} & A_{32} & \dots & A_{n2} \\ A_{13} & A_{23} & A_{33} & \dots & A_{n3} \\ \vdots & \vdots & \vdots & & \vdots \\ A_{1n} & A_{2n} & A_{3n} & \dots & A_{nn} \end{bmatrix} \tag{50}$$

Thus the procedure for obtaining the inverse of $\mathbf{A}$ is

(1) replace each element of $\mathbf{A}$ with its cofactor $A_{ij}$,

(2) transpose the resulting matrix, and

(3) divide each element in the result by $|\mathbf{A}|$.

A few important properties of the inverse should be pointed out:

$$(\mathbf{A}\,\mathbf{B})^{-1} = \mathbf{B}^{-1}\,\mathbf{A}^{-1} \tag{51}$$

$$(\mathbf{A}^{-1})^{-1} = \mathbf{A} \tag{52}$$

$$(\mathbf{A}^{t})^{-1} = (\mathbf{A}^{-1})^{t} \tag{53}$$

where $\mathbf{A}$ and $\mathbf{B}$ are $n \times n$ matrices. We noted previously that if

$$\mathbf{A}\,\mathbf{B} = {}_n\mathbf{0}_n \tag{54}$$

it is not necessarily true that $\mathbf{A}$ or $\mathbf{B}$ are zero matrices. If, however, either $\mathbf{A}$ or $\mathbf{B}$ is nonsingular, then the other must be a zero matrix since if $\mathbf{A}$ is nonsingular, premultiplying both sides of (54) by $\mathbf{A}^{-1}$ yields

$$\begin{aligned} \mathbf{B} &= \mathbf{A}^{-1}\,{}_n\mathbf{0}_n \\ &= {}_n\mathbf{0}_n \end{aligned} \tag{55}$$

The special case of the inverse of a $2 \times 2$ matrix

$$A = \begin{bmatrix} a_{11} & a_{12} \\ a_{21} & a_{22} \end{bmatrix} \tag{56}$$

is encountered frequently enough to be committed to memory:

$$\mathbf{A}^{-1} = \frac{1}{(a_{11}\,a_{22} - a_{12}\,a_{21})} \begin{bmatrix} a_{22} & -a_{12} \\ -a_{21} & a_{11} \end{bmatrix} \tag{57}$$

The set of $n$ equations in $n$ unknowns in (2) can be solved for $x_1, x_x, \dots, x_n$ with the inverse matrix. Writing (2) in matrix form as

$$\mathbf{A}\,\mathbf{X} = \mathbf{B} \tag{58}$$

we premultiply (58) on both sides by $\mathbf{A}^{-1}$ to yield

$$\mathbf{X} = \mathbf{A}^{-1}\,\mathbf{B} \tag{59}$$

*Example 1-5*

Solve the set of three equations in Example 1-1 with the matrix inverse.

*Solution*

Writing the equations in matrix form as in Example 1-4 we first compute the inverse of

$$\mathbf{A} = \begin{bmatrix} 2 & 1 & -1 \\ 3 & 2 & 1 \\ 1 & 1 & -2 \end{bmatrix}$$

the determinant was found to be

$$|\mathbf{A}| = -4$$

Replace each element of $\mathbf{A}$ with its cofactor, transpose the result and divide each element by $|\mathbf{A}|$:

$$\mathbf{A}^{-1} = \frac{1}{-4}\begin{bmatrix} 2 & 1 & 1 & -1 & 1 & -1 \\ 1 & -2 & 1 & -2 & 2 & 1 \\ 3 & 1 & 2 & -1 & 2 & -1 \\ 1 & -2 & 1 & -2 & 3 & 1 \\ 3 & 2 & 2 & 1 & 2 & 1 \\ 1 & 1 & 1 & 1 & 3 & 2 \end{bmatrix}$$

$$= \begin{bmatrix} 5/4 & -1/4 & -3/4 \\ -7/4 & 3/4 & 5/4 \\ -1/4 & 1/4 & -1/4 \end{bmatrix}$$

and

$$\mathbf{X} = \mathbf{A}^{-1}\,\mathbf{B}$$

or

$$\begin{bmatrix} x_1 \\ x_2 \\ x_3 \end{bmatrix} = \begin{bmatrix} 5/4 & -1/4 & -3/4 \\ -7/4 & 3/4 & 5/4 \\ -1/4 & 1/4 & -1/4 \end{bmatrix} \begin{bmatrix} 4 \\ 2 \\ 1 \end{bmatrix}$$

Carrying out the multiplication $\mathbf{A}^{-1}\,\mathbf{B}$ we obtain

$$x_1 = \frac{5}{4}(4) - {}^1\!/_4\,(2) - \frac{3}{4}\,(1)$$

$$= \frac{15}{4}$$

$$x_2 = \frac{7}{4}(4) + \frac{3}{4}(2) + \frac{5}{4}\,(1)$$

$$= -\frac{17}{4}$$

$$x_3 = -{}^1\!/_4\,(4) + {}^1\!/_4\,(2) - {}^1\!/_4\,(1)$$

$$= -\frac{3}{4}$$

## Solved Problems

*Gauss Elimination*

*1.1* For the resistive electric circuit shown in Fig. 1.2 the mesh currents $I_1, I_2$ and $I_3$ are defined as shown. The equations are:

$$\textit{mesh } 1\text{: } 4I_1 - 2I_2 - I_3 = 0$$

$$\textit{mesh } 2\text{: } -2I_1 + 6I_2 - 3I_3 = 6$$

$$\textit{mesh } 3\text{: } -1I_1 - 3I_2 + 6I_3 = 9$$

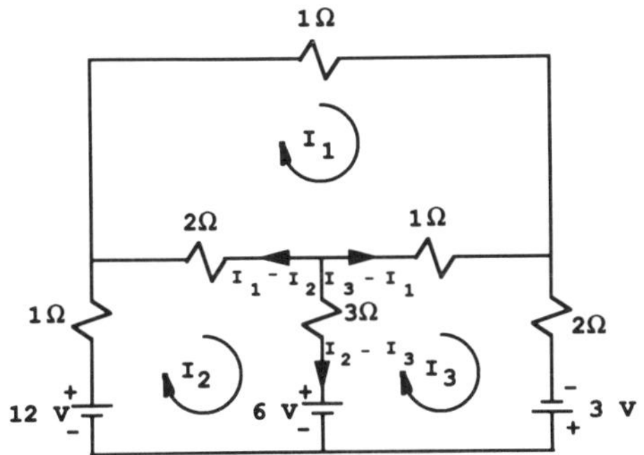


Figure 1.2

Solve these equations for $I_1, I_2$ and $I_3$ and compute the power delivered to the 3 Ω resistor.

*Solution*

Triangularization:

Step 1:

$$4I_1 - 2I_2 - I_3 = 0$$

$$0 + 5I_2 - \frac{7}{2}I_3 = 6$$

$$0 - \frac{7}{2}I_2 + \frac{23}{4}I_3 = 9$$

Step 2:

$$4I_1 - 2I_2 - I_3 = 0$$

$$5I_2 - \frac{7}{2}I_3 = 6$$

$$\frac{66}{20}I_3 = \frac{132}{10}$$

Back Substitution:

$$I_3 = \frac{44}{11} \ A$$

$$I_2 = \frac{44}{11} \ A$$

$$I_1 = 3 \ A$$

The power delivered to the 3Ω resistor is

$$P_{3\Omega} = 3\,(I_2 - I_3)^2$$

$$= 0\ W$$

*1.2* Two bodies $A$ and $B$ weighing 200*lb.* and 50*lb.*, respectively, are held in equilibrium on smooth rods by a connecting, flexible cable that makes and angle θ with the horizontal as shown in Fig. 1.3(a). Find the forces of the rods on the bodies, the tensile stress in the cable, and the angle θ.

*Solution*

The free-body diagrams are shown in Fig. 1.3(b). Applying the equations of equilibrium we have,

$$For\ A: \quad \Sigma F_x = T\cos\theta - R_A \sin 30^o = 0$$

$$\Sigma F_y = T\sin\theta + R_A \cos 30^o - 200 = 0$$

$$\text{For } B: \quad \Sigma F_x = -T \cos\theta + R_B \cos 30^o = 0$$

$$\Sigma F_y = -T \sin\theta + R_B \sin 30^o - 50 = 0$$

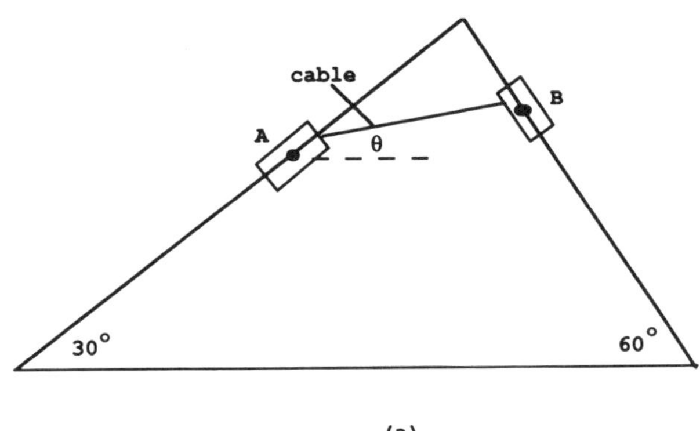


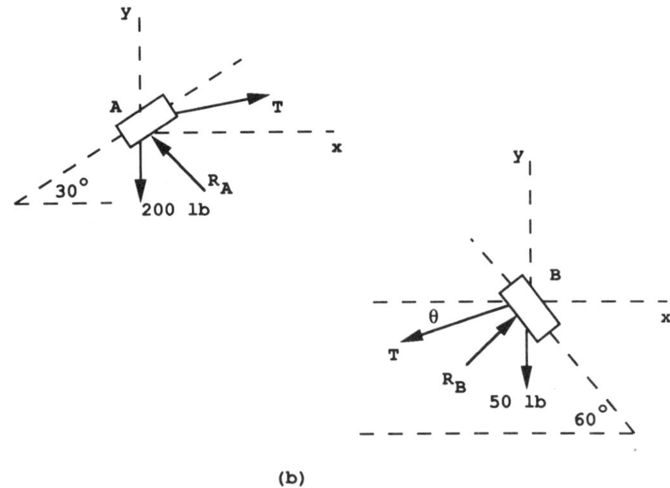


Figure 1.3

Defining $T_c = T \cos\theta$ and $T_s = T \sin\theta$ we have four equations in $T_c, T_s, R_A$ and $R_B$:

$$T_c - \sin 30^o \, R_A = 0$$

$$T_s + \cos 30^o \, R_A = 200$$

$$-T_c + \cos 30^o \, R_B = 0$$

$$-T_s + \sin 30^o \, R_B = 50$$

Adding the first equation to the third and the second equation to the fourth yields

$$T_c - \sin 30^o \, R_A = 0$$

$$T_s + \cos 30^o \, R_A = 200$$

$$-\sin 30^o \, R_A + \cos 30^o \, R_B = 0$$

$$\cos 30^o \, R_A + \sin 30^o \, R_B = 250$$

Now multiplying the third equation by $\cos 30^o / \sin 30^o$ and adding to the fourth yields

$$T_c - \sin 30^o \, R_A = 0$$

$$T_s + \cos 30^o \, R_A = 200$$

$$-\sin 30^o \, R_A + \cos 30^o \, R_B = 0$$

$$\left[\frac{\cos^2 30^o}{\sin 30^o} + \sin 30^o\right] R_B = 250$$

By back substitution:

$$R_B = 250 \sin 30^o$$

$$= 125 \; lb.$$

$$R_A = \frac{\cos 30^o}{\sin 30^o} R_B$$

$$= 216.5 \; lb.$$

$$T_s = 200 - \cos 30^o \, R_A$$

$$= 12.5$$

$$T_c = \sin 30^o \, R_A$$

$$= 108.26$$

Squaring $T_s = T \sin\theta$ and $T_c = T \cos\theta$ and adding gives

$$T^2 \sin^2\theta + T^2 \cos^2\theta = (12.5)^2 + (108.26)^2$$

and

$$T = 109 \ lb.$$

and

$$\theta = 6.59^o$$

*1.3* Solve the following set of equations by Gauss elimination:

$$x + 2y + z = 4$$

$$2x + y - z = 5$$

$$x - y + 2z = -3$$

*Solution*

Multiply the first equation by –2 and add to the second equation, and multiply the first equation by –1 and add to the third equation:

$$x + 2y + z = 4$$

$$-3y - 3z = -3$$

$$-3y + z = -7$$

Subtract the second equation from the third:

$$x + 2y + z = 4$$

$$-3y - 3z = -3$$

$$4z = -4$$

By back substitution

$$z = -1$$

$$y = 2$$

$$x = 1$$

*Matrix Algebra*

*1.4* For the following $2 \times 2$ matrices

$$\mathbf{A} = \begin{bmatrix} 1 & 1 \\ 2 & 1 \end{bmatrix} \quad \mathbf{B} = \begin{bmatrix} 1 & -1 \\ 1 & 0 \end{bmatrix}$$

show that

$$(\mathbf{A} - 2\mathbf{B})(\mathbf{A} - 3\mathbf{B}) \neq \mathbf{A}^2 - 5\,\mathbf{A}\,\mathbf{B} + 6\mathbf{B}^2$$

Determine conditions on **A** and **B** for this to be true.

*Solution*

$$\mathbf{A} - 2\mathbf{B} = \begin{bmatrix} -1 & 3 \\ 0 & 1 \end{bmatrix}$$

$$\mathbf{A} - 3\mathbf{B} = \begin{bmatrix} -2 & 4 \\ -1 & 1 \end{bmatrix}$$

$$(\mathbf{A} - 2\mathbf{B})(\mathbf{A} - 3\mathbf{B}) = \begin{bmatrix} -1 & 3 \\ 0 & 1 \end{bmatrix} \begin{bmatrix} -2 & 4 \\ -1 & 1 \end{bmatrix}$$

$$= \begin{bmatrix} -1 & -1 \\ -1 & 1 \end{bmatrix}$$

$$\mathbf{A}^2 = \begin{bmatrix} 1 & 1 \\ 2 & 1 \end{bmatrix} \begin{bmatrix} 1 & 1 \\ 2 & 1 \end{bmatrix} = \begin{bmatrix} 3 & 2 \\ 4 & 3 \end{bmatrix}$$

$$\mathbf{AB} = \begin{bmatrix} 1 & 1 \\ 2 & 1 \end{bmatrix} \begin{bmatrix} 1 & -1 \\ 1 & 0 \end{bmatrix} = \begin{bmatrix} 2 & -1 \\ 3 & -2 \end{bmatrix}$$

$$\mathbf{B}^2 = \begin{bmatrix} 1 & -1 \\ 1 & 0 \end{bmatrix} \begin{bmatrix} 1 & -1 \\ 1 & 0 \end{bmatrix} = \begin{bmatrix} 0 & -1 \\ 1 & -1 \end{bmatrix}$$

$$\mathbf{A}^2 - 5\mathbf{AB} + 6\mathbf{B}^2 = \begin{bmatrix} -7 & 1 \\ -5 & 7 \end{bmatrix}$$

Clearly

$$(\mathbf{A} - 2\mathbf{B})(\mathbf{A} - 3\mathbf{B}) = \mathbf{A}^2 - 2\mathbf{BA} - 3\mathbf{AB} + 6\mathbf{B}^2$$

(Check it anyway.) So if

$$\mathbf{BA} = \mathbf{AB}$$

then

$$(\mathbf{A} - 2\mathbf{B})(\mathbf{A} - 3\mathbf{B}) = \mathbf{A}^2 - 5\mathbf{AB} + 6\mathbf{B}^2$$

For example, the reader should show that this is true if

$$\mathbf{A} = \begin{bmatrix} 2 & 1 \\ 1 & 2 \end{bmatrix} \qquad \mathbf{B} = \begin{bmatrix} 1 & 2 \\ 2 & 1 \end{bmatrix}$$

*1.5* A electrical two-port shown in Fig. 1.4(a) is defined by its port voltages $V_1$ and $V_2$ and port currents $I_1$ and $I_2$. This two-port may represent a transistor amplifier, an automatic control system, or individual components such as a transformer. The port variables may be related in several ways. A useful representation is the chain parameter or *ABCD* matrix:

$$\begin{bmatrix} V_2 \\ I_2 \end{bmatrix} = \begin{bmatrix} A & B \\ C & D \end{bmatrix} \begin{bmatrix} V_1 \\ I_1 \end{bmatrix}$$

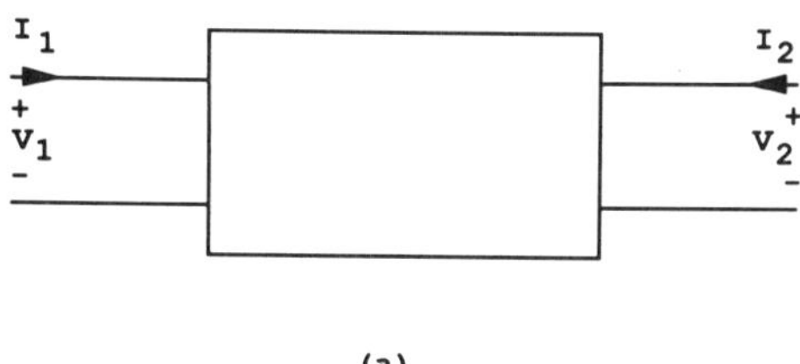


(a)

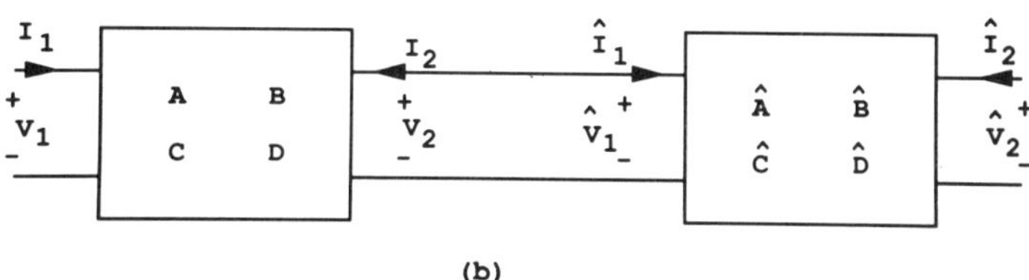


(b)

Figure 1.4

Suppose two such two-ports are cascaded as shown in Fig. 1.4(b). Determine the overall chain parameter matrix of the cascade.

*Solution*

For each two-port

$$\begin{bmatrix} V_2 \\ I_2 \end{bmatrix} = \begin{bmatrix} A & B \\ C & D \end{bmatrix} \begin{bmatrix} V_1 \\ I_1 \end{bmatrix}$$

$$\begin{bmatrix} \hat{V}_2 \\ \hat{I}_2 \end{bmatrix} = \begin{bmatrix} \hat{A} & \hat{B} \\ \hat{C} & \hat{D} \end{bmatrix} \begin{bmatrix} \hat{V}_1 \\ \hat{I}_1 \end{bmatrix}$$

But $V_2 = \hat{V}_1$, $I_2 = -\hat{I}_1$ or

$$\begin{bmatrix} \hat{V}_1 \\ \hat{I}_1 \end{bmatrix} = \begin{bmatrix} 1 & 0 \\ 0 & -1 \end{bmatrix} \begin{bmatrix} V_2 \\ I_2 \end{bmatrix}$$

So the overall chain parameter matrix of the cascade is

$$\begin{bmatrix} \hat{V}_2 \\ \hat{I}_2 \end{bmatrix} = \begin{bmatrix} \hat{A} & \hat{B} \\ \hat{C} & \hat{D} \end{bmatrix} \begin{bmatrix} \hat{V}_1 \\ \hat{I}_1 \end{bmatrix}$$

$$= \begin{bmatrix} \hat{A} & \hat{B} \\ \hat{C} & \hat{D} \end{bmatrix} \begin{bmatrix} 1 & 0 \\ 0 & -1 \end{bmatrix} \begin{bmatrix} V_2 \\ I_2 \end{bmatrix}$$

$$= \begin{bmatrix} \hat{A} & \hat{B} \\ \hat{C} & \hat{D} \end{bmatrix} \begin{bmatrix} 1 & 0 \\ 0 & -1 \end{bmatrix} \begin{bmatrix} A & B \\ C & D \end{bmatrix} \begin{bmatrix} V_1 \\ I_1 \end{bmatrix}$$

$$= \begin{bmatrix} \hat{A} & \hat{B} \\ \hat{C} & \hat{D} \end{bmatrix} \begin{bmatrix} A & B \\ -C & -D \end{bmatrix} \begin{bmatrix} V_1 \\ I_1 \end{bmatrix}$$

$$= \begin{bmatrix} (\hat{A}A - \hat{B}C) & (\hat{A}B - \hat{B}D) \\ (\hat{C}A - \hat{D}C) & (\hat{C}B - \hat{D}D) \end{bmatrix} \begin{bmatrix} V_1 \\ I_1 \end{bmatrix}$$

*Determinants*

*1.6* Using the Laplace expansion rule show the following properties of a determinant:

(a) Interchanging two rows or columns changes the sign of a determinant.

(b) If two rows or columns of a determinant are equal, the value is zero.

(c) Multiplying a row or column by a scalar $k$ multiplies the value of the determinant by $k$.

(d) Replacement of a row (column) by the sum of that row (column) and the product of $k$ times another row (column) leaves the determinant unchanged.

*Solution*

We will demonstrate these properties using the $3 \times 3$ determinant

$$|A| = \begin{vmatrix} a_{11} & a_{12} & a_{13} \\ a_{21} & a_{22} & a_{23} \\ a_{31} & a_{32} & a_{33} \end{vmatrix}$$

The extension to general $n \times n$ determinants will then become obvious. Property (a) states that, for example, for

$$|\hat{A}| = \begin{vmatrix} a_{21} & a_{22} & a_{23} \\ a_{11} & a_{12} & a_{13} \\ a_{31} & a_{32} & a_{33} \end{vmatrix}$$

we have

$$|\hat{A}| = -|A|$$

If we choose to expand $|A|$ along the first row we obtain

$$|A| = a_{11}M_{11} - a_{12}M_{12} + a_{13}M_{13}$$
$$= a_{11}A_{11} + a_{12}A_{12} + a_{13}A_{13}$$

Expanding the determinant formed by interchanging rows 1 and 2 along the second row gives

$$|\hat{A}| = -a_{11}\hat{M}_{12} + a_{12}\hat{M}_{22} - a_{13}\hat{M}_{23}$$
$$= -|A|$$

since $\hat{M}_{12} = M_{11}$, $\hat{M}_{22} = M_{12}$, $\hat{M}_{23} = M_{13}$. It should now be clear that for a general $n \times n$ determinant, we obtain the same result because interchanging rows and always expanding along the row having the same elements as before the interchange merely changes the signs of the cofactors.

To prove property (b) we investigate

$$|A| = \begin{vmatrix} a_{11} & a_{12} & a_{13} \\ a_{11} & a_{12} & a_{13} \\ a_{31} & a_{32} & a_{33} \end{vmatrix}$$

Expanding along the first row we obtain the negative of the expansion along the second row. But these must be equal since it should not matter which column or row is expanded along. Clearly the only logical value is zero. This is also easy to see if we expand along the third row of $|A|$: all the cofactors are zero.

As for property (c) we investigate

$$|A| = \begin{vmatrix} ka_{11} & ka_{12} & ka_{13} \\ a_{21} & a_{22} & a_{23} \\ a_{31} & a_{32} & a_{33} \end{vmatrix}$$

Expanding along the row which as been multiplied by $k$ it is easy to see that $k$ multiplies each factor in the result.

For property (d) we multiply the second row by $k$, add to the first row and replace the first row with the result:

$$|A| = \begin{vmatrix} (a_{11}+ka_{21}) & (a_{12}+ka_{22}) & (a_{13}+ka_{23}) \\ a_{21} & a_{22} & a_{23} \\ a_{31} & a_{32} & a_{33} \end{vmatrix}$$

Expanding along the first row we obtain

$$|A| = (a_{11}+ka_{21})\,A_{11} + (a_{12}+ka_{22})\,A_{12} + (a_{13}+ka_{23})\,A_{13}$$

Factoring this result we obtain

$$|A| = a_{11}\,A_{11} + a_{12}\,A_{12} + a_{13}\,A_{13}$$

$$+\,k\,(a_{21}\,A_{11} + a_{22}\,A_{12} + a_{23}\,A_{13})$$

But $a_{21}A_{11} + a_{22}A_{12} + a_{23}A_{13} = 0$ since this is the determinant of the matrix with rows 1 and 2 identical.

*1.7* Compute the determinant of the following $3 \times 3$ tridiagonal matrix:

$$\mathbf{A} = \begin{bmatrix} x_1 & 1 & 0 \\ 1 & x_2 & 1 \\ 0 & 1 & x_3 \end{bmatrix}$$

*Solution*

By Laplace expansion, expanding along the first row,

$$|A| = x_1 \begin{vmatrix} x_2 & 1 \\ 1 & x_3 \end{vmatrix} - 1 \begin{vmatrix} 1 & 1 \\ 0 & x_3 \end{vmatrix}$$

$$= x_1\,x_2\,x_3 - x_1 - x_3$$

By the operations in Problem 1.6, multiplying the first row by $-1/x_1$ and adding to the second:

$$|A| = \begin{vmatrix} x_1 & 1 & 0 \\ 0 & (x_2 - \dfrac{1}{x_1}) & 1 \\ 0 & 1 & x_3 \end{vmatrix}$$

Multiplying the second row by $-1/(x_2 - 1/x_1)$ and adding to the third:

$$|A| = \begin{vmatrix} x_1 & 1 & 0 \\ 0 & (x_2 - \dfrac{1}{x_1}) & 1 \\ 0 & 0 & [x_3 - \dfrac{1}{(x_2 - \dfrac{1}{x_1})}] \end{vmatrix}$$

Now the determinant of this is simply the product of the main diagonal terms:

$$|A| = x_1\,(x_2 - \frac{1}{x_1})\left[x_3 - \frac{1}{(x_2 - \frac{1}{x_1})}\right]$$

$$= x_1\left[x_3\,(x_2 - \frac{1}{x_1}) - 1\right]$$

$$= x_1\,x_2\,x_3 - x_3 - x_1$$

*Cramer's Rule*

*1.8* Using the properties of determinants obtained in Problem 1.6 prove Cramer's rule.

*Solution*

We will show this for a set of three equations and the generalization to $n$ equations will be obvious. Consider

$$\begin{bmatrix} a_{11} & a_{12} & a_{13} \\ a_{21} & a_{22} & a_{23} \\ a_{31} & a_{32} & a_{33} \end{bmatrix}\begin{bmatrix} x_1 \\ x_2 \\ x_3 \end{bmatrix} = \begin{bmatrix} b_1 \\ b_2 \\ b_3 \end{bmatrix}$$

What we must show is that

$$x_1 = \frac{\begin{vmatrix} b_1 & a_{12} & a_{13} \\ b_2 & a_{22} & a_{23} \\ b_3 & a_{32} & a_{33} \end{vmatrix}}{|A|}$$

$$x_2 = \frac{\begin{vmatrix} a_{11} & b_1 & a_{13} \\ a_{21} & b_2 & a_{23} \\ a_{31} & b_3 & a_{33} \end{vmatrix}}{|A|}$$

$$x_3 = \frac{\begin{vmatrix} a_{11} & a_{12} & b_1 \\ a_{21} & a_{22} & b_2 \\ a_{31} & a_{32} & b_3 \end{vmatrix}}{|A|}$$

satisfy the equations. Consider the first equation:

$$a_{11}x_1 + a_{12}x_2 + a_{13}x_3 = b_1$$

Substituting the assumed solutions we obtain

$$\frac{a_{11}\begin{vmatrix} b_1 & a_{12} & a_{13} \\ b_2 & a_{22} & a_{23} \\ b_3 & a_{32} & a_{33} \end{vmatrix} + a_{12}\begin{vmatrix} a_{11} & b_1 & a_{13} \\ a_{21} & b_2 & a_{23} \\ a_{31} & b_3 & a_{33} \end{vmatrix} + a_{13}\begin{vmatrix} a_{11} & a_{12} & b_1 \\ a_{21} & a_{22} & b_2 \\ a_{31} & a_{32} & b_3 \end{vmatrix}}{|A|} \stackrel{?}{=} b_1$$

This becomes

$$\frac{a_{11}(b_1A_{11} + b_2A_{21} + b_3A_{31}) + a_{12}(b_1A_{12} + b_2A_{22} + b_3A_{32}) + a_{13}(b_1A_{13} + b_2A_{23} + b_3A_{33})}{|A|} \stackrel{?}{=} b_1$$

Grouping terms we have

$$\frac{b_1\overbrace{(a_{11}A_{11} + a_{12}A_{12} + a_{13}A_{13})}^{|A|} + b_2\overbrace{(a_{11}A_{12} + a_{12}A_{22} + a_{13}A_{23})}^{0} + b_3\overbrace{(a_{11}A_{31} + a_{12}A_{32} + a_{13}A_{33})}^{0}}{|A|} \stackrel{?}{=} b_1$$

Note that the coefficients of $b_2$ and $b_3$ are zero since these represent determinants, each of which have two identical rows. For example,

$$(a_{11}A_{21} + a_{12}A_{22} + a_{13}A_{23}) = \begin{vmatrix} a_{11} & a_{12} & a_{13} \\ a_{11} & a_{12} & a_{13} \\ a_{31} & a_{32} & a_{33} \end{vmatrix}$$

$$= 0$$

The second and third equations of the set can be similarly shown to be satisfied by the $x_1, x_2, x_3$ obtained from Cramer's rule.

*1.9* Solve Problem 1.1 with Cramer's rule. Compare the effort required to that of Gauss elimination.

*Solution*

$$\begin{matrix} 4 & -2 & -1 \\ -2 & 6 & -3 \\ -1 & -3 & 6 \end{matrix} = 4\begin{vmatrix} 6 & -3 \\ -3 & 6 \end{vmatrix} + 2\begin{vmatrix} -2 & -3 \\ -1 & 6 \end{vmatrix} -1\begin{vmatrix} -2 & 6 \\ -1 & -3 \end{vmatrix}$$

$$= 4\,(36 - 9) + 2\,(-12 - 3) -1\,(6 + 6)$$

$$= 66$$

$$I_1 = \frac{\begin{vmatrix} 0 & -2 & -1 \\ 6 & 6 & -3 \\ 9 & 3 & 6 \end{vmatrix}}{66}$$

$$= \frac{2\begin{vmatrix} 6 & -3 \\ 9 & 6 \end{vmatrix} -1\begin{vmatrix} 6 & 6 \\ 9 & -3 \end{vmatrix}}{66}$$

$$= \frac{198}{66}$$

$$= 3$$

$$I_2 = \frac{\begin{vmatrix} 4 & 0 & -1 \\ -2 & 6 & -3 \\ -1 & 9 & 6 \end{vmatrix}}{66}$$

$$= \frac{4\begin{vmatrix} 6 & -3 \\ 9 & 6 \end{vmatrix} -1\begin{vmatrix} -2 & 6 \\ -1 & 9 \end{vmatrix}}{66}$$

$$= \frac{264}{66}$$

$$= \frac{44}{11}$$

$$I_3 = \frac{\begin{vmatrix} 4 & -2 & 0 \\ -2 & 6 & 6 \\ -1 & 3 & 9 \end{vmatrix}}{66}$$

$$= \frac{4\begin{vmatrix} 6 & 6 \\ -3 & 9 \end{vmatrix} + 2\begin{vmatrix} -2 & 6 \\ -1 & 9 \end{vmatrix}}{66}$$

$$= \frac{264}{66}$$

$$= \frac{44}{11}$$

*Matrix Inversion*

*1.10* Solve the following "tridiagonal" set of equations by Gauss elimination, Cramer's rule and the matrix inversion:

$$x_1 + x_2 = 1$$

$$x_1 + x_2 + x_3 = 1$$

$$x_2 + x_3 = 1$$

*Solution*

By Gauss elimination we obtain by subtracting row 1 from row 2

$$x_1 + x_2 = 1$$

$$x_3 = 0$$

$$x_2 + x_3 = 1$$

We need not carry out the triangularization any further. The result is

$$x_3 = 0$$

$$x_2 = 1$$

$$x_1 = 0$$

In matrix form

$$\begin{bmatrix} 1 & 1 & 0 \\ 1 & 1 & 1 \\ 0 & 1 & 1 \end{bmatrix} \begin{bmatrix} x_1 \\ x_2 \\ x_3 \end{bmatrix} = \begin{bmatrix} 1 \\ 1 \\ 1 \end{bmatrix}$$

**A** **X** **B**

By Cramer's rule

$$|\mathbf{A}| = \begin{vmatrix} 1 & 1 & 0 \\ 1 & 1 & 1 \\ 0 & 1 & 1 \end{vmatrix}$$

$$= -1$$

and

$$x_1 = \frac{\begin{vmatrix} 1 & 1 & 0 \\ 1 & 1 & 1 \\ 1 & 1 & 1 \end{vmatrix}}{|\mathbf{A}|} = 0$$

$$x_2 = \frac{\begin{vmatrix} 1 & 1 & 0 \\ 1 & 1 & 1 \\ 0 & 1 & 1 \end{vmatrix}}{|\mathbf{A}|} = \frac{|\mathbf{A}|}{|\mathbf{A}|} = 1$$

$$x_3 = \frac{\begin{vmatrix} 1 & 1 & 1 \\ 1 & 1 & 1 \\ 0 & 1 & 1 \end{vmatrix}}{|\mathbf{A}|} = 0$$

By matrix inversion

$$\mathbf{A}^{-1} = \frac{1}{|\mathbf{A}|} \begin{bmatrix} 0 & -1 & 1 \\ -1 & 1 & -1 \\ 1 & -1 & 0 \end{bmatrix}$$

$$= \begin{bmatrix} 0 & 1 & -1 \\ 1 & -1 & 1 \\ -1 & 1 & 0 \end{bmatrix}$$

and

$$\begin{bmatrix} x_1 \\ x_2 \\ x_3 \end{bmatrix} = \begin{bmatrix} 0 & 1 & -1 \\ 1 & -1 & 1 \\ -1 & 1 & 0 \end{bmatrix} \begin{bmatrix} 1 \\ 1 \\ 1 \end{bmatrix}$$

$$\mathbf{X} \qquad \mathbf{A}^{-1} \qquad \mathbf{B}$$

$$= \begin{bmatrix} 0 \\ 1 \\ 0 \end{bmatrix}$$

*1.11* Find a pair of square, nonzero matrices such that their product is the zero matrix as in (54).

*Solution*

Suppose $\mathbf{A}$ and $\mathbf{B}$ are nonsingular, $n \times n$ matrices. In order for $\mathbf{A}\,\mathbf{B} = {}_n\mathbf{0}_n$ we must have

$$\mathbf{A} = {}_n\mathbf{0}_n\,\mathbf{B}^{-1} = {}_n\mathbf{0}_n$$

and

$$\mathbf{B} = \mathbf{A}^{-1}\,{}_n\mathbf{0}_n = {}_n\mathbf{0}_n$$

Thus either $\mathbf{A}$ or $\mathbf{B}$ must be singular in order for $\mathbf{A}\,\mathbf{B} = {}_n\mathbf{0}_n$. Choose

$$\mathbf{A} = \begin{bmatrix} 1 & 1 \\ 1 & 1 \end{bmatrix}$$

and

$$\mathbf{B} = \begin{bmatrix} 1 & 1 \\ -1 & -1 \end{bmatrix}$$

*1.12* Show that

(*a*) $(\mathbf{A}\,\mathbf{B})^{-1} = \mathbf{B}^{-1}\,\mathbf{A}^{-1}$

(*b*) $(\mathbf{A}^{-1})^{-1} = \mathbf{A}$

(*c*) $(\mathbf{A}^t)^{-1} = (\mathbf{A}^{-1})^t$

*Solution*

To show property (a) we premultiply both sides by $\mathbf{A}\,\mathbf{B}$:

$$(\mathbf{A}\,\mathbf{B})\,(\mathbf{A}\,\mathbf{B})^{-1} = \mathbf{A}\,\mathbf{B}\,\mathbf{B}^{-1}\,\mathbf{A}^{-1}$$

$$= \mathbf{1}_n$$

To show (b) we take the inverse of both sides. To show (c) we note that the form of $\mathbf{A}^{-1}$ given in (50). Transposing (50) gives

$$(\mathbf{A}^{-1})^t = \frac{1}{|\mathbf{A}|}\begin{bmatrix} A_{11} & A_{12} & A_{13} & \dots & A_{1n} \\ A_{21} & A_{22} & A_{23} & \dots & A_{2n} \\ \vdots & & & & \vdots \\ A_{n1} & A_{n2} & \dots & & A_{nn} \end{bmatrix}$$

But it is easy to see that the inverse of $\mathbf{A}^t$ is identically equal to this.

CHAPTER

2

# ORDINARY DIFFERENTIAL EQUATIONS

## 2.1 The Linear, Constant-Coefficient, Ordinary Differential Equation

The most frequently encountered ordinary differential equation in the study of linear, dynamical systems is the *n-th order, linear, constant-coefficient, ordinary, differential equation*:

$$\frac{d^n y(t)}{dt^n} + a_{n-1} \frac{d^{n-1} y(t)}{dt^{n-1}} + \dots + a_1 \frac{dy(t)}{dt} + a_0 y(t) = f(t) \tag{1}$$

The variable $y$ is said to be the *dependent variable* and $t$ is the *independent variable.* The function $f(t)$ is called the *forcing function.* Note that $a_i$ is the coefficient of the *$i$-th derivative* of $y(t)$. Also note that, without loss of generality, we have assumed the leading coefficient, $a_n$, to be unity; if it is not, divide both sides of the equation by $a_n$ to produce the form in (1). The object is given $a_{n-1}$, ..., $a_1$, $a_0$ and f(t), find a solution, $y(t)$, to this equation.

The term *constant coefficient* means that the $a_i$ coefficients are not functions of the independent variable, $t$. The term *ordinary* is simply used to distinguish these from partial differential equations discussed in the next chapter in the sense that only ordinary derivatives, as opposed to partial derivatives, appear in (1). Suppose $y$ is a function of two variables, $x$ and $t$, as $y(x,t)$. A partial differential equation would involve partial derivatives such as

$$a_{23} \frac{\partial^2 y(x,t)}{\partial x^2} + a_{22} \frac{\partial^2 y(x,t)}{\partial x \, \partial t} + a_{21} \frac{\partial^2 y(x,t)}{\partial t^2} \tag{2}$$

$$+ a_{12} \frac{\partial y(x,t)}{\partial x} + a_{11} \frac{\partial y(x,t)}{\partial t} + a_0 \, y(x,t) = f(x,t)$$

The equation in (1) is *n-th order* in that the highest derivative appearing in the equation is the $n$-th derivative. The term *linear* means that neither $y(t)$ nor any of its derivatives are raised to a power greater than one and $y(t)$ and/or its derivatives do not appear as products. The following ordinary differential equations are nonlinear:

$$\frac{d^2y(t)}{dt^2} \;\; 2y^3(t) = 2t \tag{3}$$

$$\left[ \frac{dy(t)}{dt} \right]^2 + 3y(t) = 2 + 4t^2 \tag{4}$$

$$\frac{d^2y(t)}{dt^2} + 2y(t) \frac{dy(t)}{dt} + 6y(t) = [\cos\,(2t)]^2 \tag{5}$$

The terms $y^3(t)$ in (3), $\left[ \frac{dy(t)}{dt} \right]^2$ in (4) and $2y(t) \frac{dy(t)}{dy}$ in (5) make each of these differential equations nonlinear. Note that $\frac{d^2y(t)}{dt^2}$ in (3), $4t^2$ in (4) and $[cos\,(2t)]^2$ in (5) do not cause the respective equations to be nonlinear. Also note that (3) is second order as is (5) but (4) is first order, i.e., there is an important difference between $\frac{d^k y(t)}{dt^k}$ and $\left[ \frac{dy(t)}{dt} \right]^k$. We will only consider the solution to linear, constant coefficient, ordinary differential equations in this chapter.

## 2.2 The Property of Superposition

The linearity of a differential equation has some very important consequences. To simplify notation, we will denote the $k$-th derivative of $y(t)$ as

$$\frac{d^k y(t)}{dt^k} \stackrel{\Delta}{=} y^k \tag{6}$$

The dependence of $y$ and $f$ on $t$ will be dropped, i.e., $y(t) \stackrel{\Delta}{=} y$ and $f(t) \stackrel{\Delta}{=} f$. Suppose $y_1$ satisfies (1) with $f(t) = f_1$:

$$y_1^{(n)} + a_{n-1} y_1^{(n-1)} + \ldots + a_0 y_1 = f_1 \tag{7a}$$

and suppose $y_2$ satisfies (1) with $f = f_2$:

$$y_2^{(n)} + a_{n-1} y_2^{(n-1)} + \ldots + a_0 y_2 = f_2 \tag{7b}$$

then it is true that $y = y_1 + y_2$ satisfies (1) when $f = f_1 + f_2$. This is called the property of superposition. Also $\beta_1 y_1$ satisfies (1) if $f = \beta_1 f_1$ as does $\beta_2 y_2$ when $f = \beta_2 f_2$. These properties of a linear differential equation can be shown by multiplying (7a) by $\beta_1$ and (7b) by $\beta_2$, adding the two equations and combining as

$$[\beta_1 y_1 + \beta_2 y_2]^{(n)} + a_{n-1} [\beta_1 y_1 + \beta_2 y_2]^{(n-1)} + \ldots + a_0 [\beta_1 y_1 + \beta_2 y_2] = [\beta_1 f_1 + \beta_2 f_2] \tag{8}$$

Consequently, the solution to

$$y^{(n)} + a_{n-1} y^{(n-1)} + \ldots + a_0 y = f_1 + f_2 + \ldots + f_k \tag{9}$$

is the sum of the solutions to the equations

$$\begin{aligned} y_1^{(n)} + a_{n-1} y_1^{(n-1)} + \ldots + a_0 y_1 &= f_1 \\ y_2^{(n)} + a_{n-1} y_2^{(n-1)} + \ldots + a_0 y_2 &= f_2 \\ y_k^{(n)} + a_{n-1} y_k^{(n-1)} + \ldots + a_0 y_k &= f_k \end{aligned} \tag{10}$$

as

$$y = y_1 + y_2 + \ldots + y_k \tag{11}$$

## 2.3 Homogeneous and Particular Solutions

The solution to (1) is unchanged if we add the "zero forcing function" to the right-hand side:

$$y^{(n)} + a_{n-1} y^{(n-1)} + \ldots + a_0 y = 0 + f(t) \tag{12}$$

Consequently, the solution will, by superposition, consist of two parts. The homogeneous solution, $y_h$, (sometimes called the natural solution or the complementary solution) will be the solution of the homogeneous equation having only the zero forcing function:

$$y_h^{(n)} + a_{n-1}\, y_h^{(n-1)} + \dots + a_0\, y_h = 0 \tag{13}$$

The particular solution (sometimes called the forced solution), $y_p$, is the solution to the inhomogeneous equation with f as the forcing function:

$$y_p^{(n)} + a_{n-1}\, y_p^{(n-1)} + \dots + a_0\, y_p = f \tag{14}$$

By superposition, the total solution is the sum of the homogeneous and particular solutions:

$$y = y_h + y_p \tag{15}$$

## 2.4 The Homogeneous Solution

A solution of the homogeneous equation in (13) which will always work is of the form

$$y_h = Ce^{pt} \tag{16}$$

where C and p are, as yet, undetermined. Substituting (16) into (13) yields

$$(p^n + a_{n-1}\, p^{n-1} + \dots + a_0)\, ce^{pt} = 0 \tag{17}$$

$$p^n + a_{n-1}\, p^{n-1} + \dots + a_0 = 0 \tag{18}$$

There are $n$ roots, $p_1, p_2, \dots p_n$, so that for n "zero forcing functions" the complete homogeneous solution is

$$y_h = C_1\, e^{p_1 t} + C_2\, e^{p_2 t} + \dots + C_n\, e^{p_n t} \tag{19}$$

Therefore, the homogeneous solution may be written as

$$y_h\,(t) = C_1\, \phi_1\,(t) + C_2\phi_2(t) + \dots + C_n\phi_n(t) \tag{20}$$

where the "eigenfunctions" $\phi_i\,(t)$ are

$$\phi_1\,(t) = e^{p_1 t},\ \phi_2\,(t) = e^{p_2 t},\ \dots,\ \phi_n(t) = e^{p_n t} \tag{21}$$

*(a) Real, repeated roots*

Suppose one of the roots of (18), $p_i$, is of multiplicity $m$, i.e., $p_i = p_{i+1} = \dots = p_{i+m-1}$, then $m$ of the terms in (19) may be combined into one term. In this case, the eigenfunctions associated with this root $p_i$ are

$$\phi_i = e^{p_i t},\ \phi_{i+1} = t\phi_i = te^{p_i t},\ \phi_{i+2} = t^2\phi_i = t^2 e^{p_i t},\ \ldots,\ \phi_{i+m-1} = t^{m-1}\phi_i = t^{m-1} e^{p_i t} \tag{22}$$

None of these eigenfunctions may be combined into a single function with only one undetermined constant: an important requirement as we shall see later when boundary conditions are applied to the total solution to evaluate $C_1, C_2, \ldots, C_n$.

*(b) Complex roots*

Suppose one of the roots of (18) is complex, $p_i = \alpha_i + j\beta_i$. (See Appendix A for a discussion of complex numbers.) Since the coefficients of (18), $a_i$, are assumed to be real numbers, there must appear another root which is the complex conjugate of $p_i$, i.e., $p_{i+1} = p_i^* = \alpha_i - j\beta_i$. In this case we may write

$$C_i\,\phi_i + C_{i+1}\,\phi_{i+1} = C_i\,e^{p_i t} + C_{i+1}\,e^{p_i^* t} = e^{\alpha_i t}\,(C_i\,e^{j\beta_i t} + C_{i+1}\,e^{-j\beta_i t}) \tag{23}$$

equivalently as

$$C_i\phi_i + C_{i+1}\phi_{i+1} = A_i\,e^{\alpha_i t}\cos\beta_i t + A_{i+1}\,e^{\alpha_i t}\sin\beta_i t \tag{24}$$

where $A_i$ and $A_{i+1}$ are undetermined constants related to $C_i$ and $C_{i+1}$ by $A_i = C_i + C_{i+1}$ and $A_{i+1} = j(C_i - C_{i+1})$. Since $C_i$ and $C_{i+1}$ are, as yet, undetermined we may as well write

$$C_i\phi_i + C_{i+1}\phi_{i+1} = C_i\,e^{\alpha_i t}\cos\beta_i t + C_{i+1}\,e^{\alpha_i t}\sin\beta_i t \tag{25}$$

so that the eigenfunctions for a complex root and its conjugate are

$$\phi_i = e^{\alpha_i t}\cos\beta_i t,\ \phi_{i+1} = e^{\alpha_i t}\sin\beta_i t \tag{26}$$

*(c) Repeated, complex roots*

Suppose a complex root $p_i$ (and its corresponding conjugate $p_i*$) is of multiplicity $m$. Then the corresponding eigenfunctions are

$$\begin{aligned} &\phi_i = e^{\alpha_i t}\cos\beta_i t, \quad \phi_{i+1} = e^{\alpha_i t}\sin\beta_i t \\ &\phi_{i+2} = t\phi_i, \qquad \phi_{i+3} = t\,\phi_{i+1} \\ &\phi_{i+4} = t^2\phi_i, \qquad \phi_{i+5} = t^2\,\phi_{i+1} \\ &\phi_{i+2m-2} = t^{m-1}\,\phi_i, \quad \phi_{i+2m-1} = t^{m-1}\,\phi_{i+1} \end{aligned} \tag{26}$$

*Example 2-1*

Suppose the roots of a 6-th order characteristic equation are

$$p_1 = 4, p_2 = 3, p_3 = 1+j2, p_4 = 1-j2, p_5 = 1+j2, p_6 = 1-j2$$

Write the form of the homogeneous solution.

*Solution*

$$y_h = C_1 e^{4t} + C_2 e^{3t} + C_3 e^t \cos 2t + C_4 e^t \sin 2t$$

$$+ C_5 te^t \cos 2t + C_6 te^t \sin 2t$$

Thus the form of the homogeneous solution is simple to obtain once the roots of the characteristic equation are found. For differential equations of order greater than 2, the factorization of the resulting characteristic equation is quite difficult and is generally accomplished with the aid of a digital computer. See Chapter 5 for a discussion of various root-finding algorithms suitable for this problem.

## 2.5 Linear Independence of Functions

A set of $n$ functions $\phi_1(t), \phi_2(t), \ldots, \phi_n(t)$ is said to be linearly independent over some interval $t_1 \le t \le t_2$ if none of the functions may be written as a linear combination of the remaining functions, i.e., it is not possible to obtain a set of constants such that

$$\phi_i(t) = C_1\phi_1(t) + \ldots + C_{i-1}\,\phi_{i-1}(t) + C_{i+1}\,\phi_{i+1}(t) \ldots + C_n\,\phi_n(t) \tag{27}$$

Mathematically, this means that there does not exist a set of constants $C_1, C_2, \ldots, C_n$, not all of which are zero, such that

$$C_1\,\phi_1(t) + C_2\,\phi_2(t) + \ldots + C_n\,\phi_n(t) = 0 \tag{28a}$$

over $t_1 \le t \le t_2$. Eigenfunctions of the homogeneous solution, $\phi_i$, chosen according to the rules given in the previous section can be shown to be linearly independent.

A simple test to determine the linear independence of a set of $n$ functions is as follows. If (28a) is true then we may differentiate (28a) $(n-1)$ times yielding

$$\begin{aligned} C_1\,\phi_1^{(1)} + C_2\,\phi_2^{(1)} + \ldots + C_n\,\phi_n^{(1)} &= 0 \\ C_1\,\phi_1^{(2)} + C_2\,\phi_2^{(2)} + \ldots + C_n\,\phi_n^{(2)} &= 0 \\ C_1\,\phi_1^{(n-1)} + C_2\,\phi_2^{(n-1)} + \ldots + C_n\,\phi_n^{(n-1)} &= 0 \end{aligned} \tag{28b}$$

Equations (28) form a set of $n$ linear, algebraic equations in the $n$ unknowns $C_1, C_2, ..., C_n$. The functions $\phi_i(t)$ will be linearly independent if the only solution of (28) is $C_1 = C_2 = ... = C_n = 0$. Utilizing Cramer's rule from the previous chapter we see that (28) will have the trivial, zero solution if and only if the following determinant is identically zero over $t_1 \le t \le t_2$:

$$W(t) = \begin{vmatrix} \phi_1 & \phi_2 & \dots & \phi_n \\ \phi_1^{(1)} & \phi_2^{(1)} & \dots & \phi_n^{(1)} \\ \phi_1^{(2)} & \phi_2^{(2)} & \dots & \phi_n^{(2)} \\ \vdots & & & \vdots \\ \phi_1^{(n-1)} & \phi_2^{(n-1)} & \dots & \phi_n^{(n-1)} \end{vmatrix} \tag{29}$$

The determinant in (29) is called the Wronskian, and its nonzero value, $W(t) \neq 0$, guarantees the linear independence of the $\phi_i(t)$ over $t_1 \le t \le t_2$.

## 2.6 The Particular Solution

The solution of the inhomogeneous equation in (14) depends on the form of $f$. The simplest technique is to assume a particular solution which is of the same *form* as $f$, substitute into (14) and determine any unknown coefficients in the form. This is known as the method of undetermined coefficients. For example, suppose

$$y^{(2)} + 2y^{(1)} + 3y = t^2$$

We assume

$$y_p = A + Bt + Ct^2$$

(Note that although $f$ involves only $t^2$ we must assume all lower order powers of $t$ in the form.) Substitution yields

$$(2C) + 2(B + 2Ct) + 3(A + Bt + Ct^2) = t^2$$

Matching powers of t we obtain

$$2C + 2B + 3A = 0$$

$$4C + 3B = 0$$

$$3C = 1$$

(Note the triangular form.) Solving by "back substitution" yields

$$C = 1/3$$

$$B = -4/9$$

$$A = 2/9$$

Note that if $f$ is decomposed into the sum of several functions:

$$f = f_1 + f_2 = \dots + f_k \tag{30}$$

Then, by superposition, the particular solution consists of the sum of the particular solutions due to each $f_i$:

$$y_p = y_{p1} + y_{p2} + \dots + y_{pk} \tag{31}$$

as illustrated by (9), (10) and (11).

The most common forms of $f$ and the assumed form of the particular solution are given in Table 2.1.

Table 2.1

| $f_i(t)$ | Form of $y_p(t)$ |
|---|---|
| $M\, t^l$ | $K_0 + K_1 t + \dots + K_l t^l$ |
| $M \cos \beta t$ | $K_c \cos \beta t + K_s \sin \beta t$ |
| $M \sin \beta t$ | $K_c \cos \beta t + K_s \sin \beta t$ |
| $M\, e^{\alpha t} \cos \beta t$ | $K_c\, e^{\alpha t} \cos \beta t + K_s\, e^{\alpha t} \sin \beta t$ |
| $M\, e^{\alpha t} \sin \beta t$ | $K_c\, e^{\alpha t} \cos \beta t + K_c\, e^{\alpha t} \sin \beta t$ |

Note that although $f$ may consist of only a cosine or only a sine, the assumed form of the particular solution *must* involve both a cosine and a sine. We may generalize this result to say that if $f_i(t)$ is of the form

$$f_i(t) = \begin{cases} M\, t^l e^{\alpha t} \cos \beta t \\ \quad or \\ M\, t^l e^{\alpha t} \sin \beta t \end{cases} \tag{32}$$

we would assume the portion of the particular solution due to this portion of $f(t)$ to be

$$y_{pi}(t) = e^{\alpha t} \cos \beta t \left[K_{c0} + K_{c1}\, t + \dots + K_{cl}\, t^l\right]$$
$$+ e^{\alpha t} \sin \beta t \left[K_{s0} + K_{s1}\, t + \dots + K_{sl}\, t^l\right] \tag{33}$$

Note the generality of the form in (32); we may have $\ell = 0$, or $\alpha = 0$, or $\beta = 0$ or any combination.

There is one situation in which these assumed forms will not work--when $f_i$ is the same form as one or more of the eigenfunctions, $\phi_j$, of the homogeneous solution. Clearly this will be a problem because $\phi_j$ satisfies the homogeneous equation in (13). Choosing a form of the particular solution which is of the same form as $f_i$ and therefore $\phi_j$ then will satisfy the homogeneous equation and cannot satisfy the inhomogeneous equation in (14). A simple way of remedying this situation is the following. Suppose $f_i$ is of the form

$$f_i(t) = \begin{cases} M\, t^l e^{\alpha t} \cos(\beta t) \\ \quad or \\ M\, t^l e^{\alpha t} \sin(\beta t) \end{cases} \tag{34}$$

If $\alpha + j\beta$ or $\alpha - j\beta$ is *not* a root of the characteristic equation choose $y_{pi}(t)$ as in (33). (This rule, of course, holds for real roots, i.e., $\beta = 0$). If $\alpha + j\beta$ or $\alpha - j\beta$ is a root of the characteristic equation of multiplicity $m$ choose $y_{pi}(t)$ to be (33) multiplied by $t^m$.

*Example 2-2*

Suppose

$$y^{(2)} + 2y^{(1)} + y = t^2 + 3\cos 2t + 2e^{-t}$$

Determine the particular solution.

*Solution*

Write

$$f = f_1 + f_2 + f_3$$

where

$$f_1 = t^2$$

$$f_2 = 3\cos 2t$$

$$f_3 = 2e^{-t}$$

For $f_1 = t^2$ we assume

$$y_{p1} = K_0 + K_1 t + K_2 t^2$$

Substitution yields

$$(2K_2) + 2(K_1 + 2K_2 t) + (K_0 + K_1 t + K_2 t^2) = t^2$$

Matching powers of $t$ we have

$$2K_2 + 2K_1 + K_0 = 0$$

$$4K_2 + K_1 = 0$$

$$K_2 = 1$$

and

$$K_2 = 1$$

$$K_1 = -4$$

$$K_0 = 6$$

For $f_2 = 3\cos 2t$ we assume

$$y_{p2} = K_{c0}\cos 2t + K_{s0}\sin 2t$$

Substitution yields

$$(-4K_{c0}\cos 2t - 4K_{s0}\sin 2t) + 2(-2K_{s0}\sin 2t + 2K_{s0}\cos 2t) +$$

$$+ (K_{c0}\cos 2t + K_{s0}\sin 2t) = 3\cos 2t$$

Matching coefficients of $\cos 2t$ and $\sin 2t$ yields

$$-3K_{c0} + 4K_{s0} = 3$$

$$-4K_{c0} - 3K_{s0} = 0$$

Solution by Cramer's rule yields

$$K_{c0} = 9/7$$

$$K_{s0} = 12/7$$

For $f_3 = 2e^{-t}$ we note that the characteristic equation has roots -1, -1 so that $\phi_1 = e^{-t}$, $\phi_2 = te^{-t}$ and -1 is a root of multiplicity m = 2. So we choose

$$y_{p3} = K\,t^2 e^{-t}$$

Substitution yields.

$$K(2e^{-t} - 2te^{-t} - 2te^{-t} + t^2e^{-t}) + 2K(2te^{-t} - t^2e^{-t}) + K(t^2e^{-t}) = 2e^{-t}$$

Matching coefficients yields K = 1. Thus the total particular solution is

$$y_p = y_{p1} + y_{p2} + y_{p3}$$

$$= 6 - 4t + t^2 + \frac{9}{7}\cos 2t + \frac{12}{7}\sin 2t + t^2 e^{-t}$$

## 2.7 The Complete Solution

The complete solution consists, by superposition, of the sum of the homogeneous and particular solutions:

$$y(t) = C_1\phi_1(t) + C_2\phi_2(t) + \dots + C_n\phi_n(t) + y_p(t) \tag{35}$$

All items except $C_1, C_2, \dots, C_n$ are known. A multitude of solutions to (1) therefore exists. To single out a solution for a specific problem, the unknown constants are obtained by applying boundary conditions determined by the problem under consideration to (35). The most common are initial conditions in which $t$ represents time and are the specification of $y(t)$ and its $n-1$ derivatives at some initial time $t_0$:

$$\begin{aligned} &y(t_0) \\ &y^{(1)}|_{t_0} \\ &\vdots \\ &y^{(n-1)}|_{t_0} \end{aligned} \tag{36}$$

We may independently specify $y$ and its first $n-1$ derivatives but equation (1) relates $y^n$ and all higher order derivatives of $y$ to these. In this case the solution holds for all $t \geq t_0$.
Thus we obtain

$$\begin{aligned} &C_1\phi_1(t_0) + C_2\phi_2(t_0) + \dots + C_n\phi_n(t_0) + y_p(t_0) = y(t_0) \\ &C_1\phi_1^{(1)}|_{t_0} + C_2\phi_2^{(1)}|_{t_0} + \dots + C_n\phi_n^{(1)}|_{t_0} + y_p^{(1)}|_{t_0} = y^{(1)}|_{t_0} \\ &\qquad\vdots \qquad\qquad\qquad \vdots \\ &C_1\phi_1^{(n-1)}|_{t_0} + C_2\phi_2^{(n-1)}|_{t_0} + \dots + C_n\phi_n^{(n-1)}|_{t_0} + y_p^{(n-1)}|_{t_0} = y^{(n-1)}|_{t_0} \end{aligned} \tag{37}$$

Equations (37) can be written in matrix form as

$$\underbrace{\begin{bmatrix} \phi_1(t_0) & \phi_2(t_0) & \dots & \phi_n(t_0) \\ \phi_1^{(1)}\big|_{t_0} & \phi_2^{(1)}\big|_{t_0} & \dots & \phi_n^{(1)}\big|_{t_0} \\ & \vdots & & \\ \phi_1^{(n-1)}\big|_{t_0} & \phi_2^{(n-1)}\big|_{t_0} & \dots & \phi_n^{(n-1)}\big|_{t_0} \end{bmatrix}}_{\mathbf{W}} \begin{bmatrix} C_1 \\ C_2 \\ \vdots \\ C_n \end{bmatrix} = \begin{bmatrix} y(t_0) - y_p(t_0) \\ y^{(1)}\big|_{t_0} - y_p^{(1)}\big|_{t_0} \\ \vdots \\ y^{(n-1)}\big|_{t_0} - y_p^{(n-1)}\big|_{t_0} \end{bmatrix} \tag{38}$$

from which $C_1, C_2, \dots, C_n$ can be determined. The coefficient matrix **W** in (38) is the Wronskian discussed in Section 2-5. The particular choices of the $\phi_1$ eigenfunctions in the manner described in Section 2-4 guarantees that the $\phi_i(t)$ are linearly independent and therefore the Wronskian is nonsingular thus guaranteeing a unique solution for $C_1, C_2, \dots, C_n$.

*Example 2-3*

Determine the solution to

$$y^{(2)} + 5y^{(1)} + 6y = 3t$$

for $t \geq 0$ subject to the initial conditions

$$y(0) = 1$$

$$y^{(1)}\big|_{t=0} = 0$$

*Solution*

The homogeneous solution is

$$y_h(t) = C_1 e^{-2t} + C_2 e^{-3t}$$

and the particular solution is

$$y_p(t) = -\frac{5}{12} + \frac{1}{2}t$$

The total solution is

$$y(t) = y_h(t) + y_p(t)$$

$$= C_1 e^{-2t} + C_2 e^{-3t} - \frac{5}{12} + \frac{1}{2}t$$

The initial conditions yield

$$y(0) = 1 = C_1 + C_2 - \frac{5}{12}$$

$$y^{(1)}\big|_{t_0} = 0 = -2C_1 - 3C_2 + 1/2$$

and we obtain

$$\begin{bmatrix} 1 & 1 \\ -2 & -3 \end{bmatrix} \begin{bmatrix} C_1 \\ C_2 \end{bmatrix} = \begin{bmatrix} \frac{17}{12} \\ -\frac{1}{2} \end{bmatrix}$$

or

$$C_1 = \frac{15}{4} \text{ and } C_2 = -\frac{7}{3}$$

Thus

$$y(t) = \frac{15}{4} e^{-2t} - \frac{7}{3} e^{-3t} - \frac{15}{12} + \tfrac{1}{2} t$$

## 2.8 The Particular Solution for Sinusoidal Forcing Functions

For the portion of the particular solution due to $f_i(t)$ of the form

$$f_i = \begin{cases} M\, e^{\alpha t} \cos\beta t & \text{(39a)} \\ \quad or \\ M\, e^{\alpha t} \sin\beta t & \text{(39b)} \end{cases}$$

we assume the particular solution to be of the form

$$y_p = K_{c0} e^{\alpha t} \cos \beta t + K_{s0}\, e^{\alpha t} \sin \beta t \tag{40}$$

substitute in (14) and determine the coefficients $K_{c0}$ and $K_{s0}$.
A much simpler technique is the following.

Instead of $f_i(t)$ given in (39) we find the particular solution, $\hat{y}_p$, with

$$f_i = M\, e^{(\alpha + j\beta)t} \tag{41}$$

If $f_i$ is the cosine form given by (39a) then

$$y_p = \text{Re}\,\{\hat{y}_p\} \tag{42}$$

and if $f_i$ is the sine form given in (39b) then

$$y_p = \text{Im}\,\{\hat{y}_p\} \tag{43}$$

where Re denotes "real part" and Im denotes "imaginary part".

To determine $\hat{y}_p$ we assume

$$\hat{y}_p = K\, e^{(\alpha + j\beta)t} \tag{44}$$

Substitution in the inhomogeneous equation:

$$y^{(n)} + a_{n-1}y^{(n-1)} + \dots + a_1 y^{(1)} + a_0 y = f_i \tag{45}$$

yields

$$[(\alpha + j\beta)^n + a_{n-1}(\alpha + j\beta)^{n-1} + \dots + a_1(\alpha + j\beta) + a_0]\, Ke^{(\alpha + j\beta)t} = Me^{(\alpha + j\beta)t} \tag{46}$$

from which we obtain

$$K = \frac{M}{(\alpha + j\beta)^n + a_{n-1}(\alpha + j\beta)^{n-1} + \dots + a_1(\alpha + j\beta) + a_0} \tag{47}$$

With complex arithmetic we obtain from (47)

$$K = |K| \angle \Theta_K \tag{48}$$

so that

$$\hat{y}_p = |K|\, e^{j\Theta_K}\, e^{\alpha t}\, e^{j\beta t} \tag{49}$$

If $f_i$ is the cosine form in (39a) we obtain

$$\begin{aligned} y_p &= \text{Re}\,\{\hat{y}_p\} \\ &= |K|\, e^{\alpha t} \cos(\beta t + \Theta_K) \end{aligned} \tag{50}$$

and if $f_i$ is the sine form in (39b) we obtain

$$\begin{aligned} y_p &= \text{Im}\,\{\hat{y}_p\} \\ &= |K|\, e^{\alpha t} \sin(\beta t + \theta_K) \end{aligned} \tag{51}$$

A simple proof of this result can be obtained by realizing that (39a) is the real part of (41) and (39b) is the imaginary part of (41). Thus (41) is the sum of (39a) and (39b) multiplied by $j$. Therefore we solve for the particular solution with $f$ given in (41) and only use the desired part of that solution, a result made possible by the linearity of the differential equation.

*Example 2-4*

Determine the particular solution to

$$y^{(2)} + 3y^{(1)} + y = 2 \cos t$$

*Solution*

By the previous method we would assume

$$y_p = K_{c0} \cos t + K_{s0} \sin t$$

and substitute:

$$[-K_{c0} \cos t - K_{s0} \sin t] + 3\,[-K_{c0} \sin t + K_{s0} \cos t] +$$
$$[K_{c0} \cos t + K_{s0} \sin t] = 2 \cos t$$

Matching coefficients we obtain $K_{s0} = 2/3$, $K_{c0} = 0$ so that

$$y_p = \frac{2}{3} \sin t$$

By the alternate method of this section we assume

$$f = 2\, e^{jt}$$
$$\hat{y}_p = K\, e^{jt}$$

and

$$K = \frac{2}{(j1)^2 + 3(j1) + 1}$$
$$= -j\, \frac{2}{3}$$

or

$$K = \frac{2}{3}\, e^{-j90^o}$$

and

$$y_p = \text{Re}\,\{\hat{y}_p\}$$
$$= \text{Re}\,\{\frac{2}{3}\, e^{\dot{\theta}t}\, e^{-j90^o}\}$$
$$= \frac{2}{3} \cos (t - 90^o)$$
$$= \frac{2}{3} \sin t$$

## 2.9 Simultaneous, Ordinary Differential Equations

Quite often in modeling a physical system we obtain not just one differential equation of the form of (1) but a simultaneous set of such equations. To simplify the following technique for solving such equations we introduce the operator D as a shorthand notation for the derivative

$$D^n y \triangleq \frac{d^n y}{dt^n} \tag{52}$$

so that $D$ "operates" on $y$ to produce the $n$-th derivative. Similarly

$$\frac{1}{D^n} y \triangleq \underbrace{\int\int\int \ldots \int}_{n \text{ integrals}} y(\tau_1, \tau_2 \ldots, \tau_n)\, d\tau_1\, d\tau_2 \ldots d\tau_n \tag{53}$$

so that

$$D^n \left\{ \frac{1}{D^n} y \right\} = y$$

The $n$-th order differential equation in (1) can be written in terms of the operator $D$ as

$$(D^n + a_{n-1} D^{n-1} + \ldots + a_1 D + a_0)\, y(t) = f(t) \tag{54}$$

Note that the operator $D$ obeys the rules of algebra.

Similarly, a simultaneous set of differential equations such as

$$\frac{dy(t)}{dt} + 2y(t) + 3\frac{dx(t)}{dt} + x(t) = f_1(t) \tag{55}$$

$$\frac{d^2y(t)}{dt^2} + \frac{dy(t)}{dt} + \frac{dx(t)}{dt} + 4x(t) = f_2(t)$$

can be written in operator notation and matrix form as

$$\begin{bmatrix} (D+2) & (3D+1) \\ (D^2+D) & (D+4) \end{bmatrix} \begin{bmatrix} y(t) \\ x(t) \end{bmatrix} = \begin{bmatrix} f_1(t) \\ f_2(t) \end{bmatrix} \tag{56}$$

In order to solve these equations we need to first eliminate $x$ and obtain one differential equation in $y$ which we may solve by the techniques of the preceding sections. Then we eliminate $y$ and obtain one differential equation in $x$ which we may solve in like fashion. To eliminate $x$ we multiply the first equation in (55) by $(D^2 + D)$ and the second by $-(D + 2)$ to obtain

$$(D^2 + D)(D + 2)y + (D^2 + D)(3D + 1)X = (D^2 + D)f_1$$

$$-(D + 2)(D^2 + D)y - (D + 2)(D + 4)X = -(D + 2)f_2 \tag{57}$$

Note that

$$(D^2 + D)(D + 2)y = (D^3 + 3D^2 + 2D)y$$

$$(D + 2)(D^2 + D)y = (D^3 + 3D^2 + 2D)y \tag{58}$$

Adding the two resulting equations gives

$$[(D^2+D)(3D+1)-(D+2)(D+4)]\,x = (D^2+D)f_1-(D+2)f_2 \tag{59}$$

or

$$[3D^3+3D^2-5D-8]\,x = D^2 f_1+Df_1-Df_2-2f_2 \tag{60}$$

Differentiating $f_1$ and $f_2$ (which are assumed known) according to the right-hand side of (60) yields one differential equation in $x$. A differential equation in $y$ may be similarly obtained.

A general set of $N$ simultaneous, linear, constant coefficient, ordinary differential equations in the variables $y_1(t), y_2(t) \ldots, y_n(t)$ can be written in matrix form as

$$\begin{bmatrix} L_{11}(D) & L_{12}(D) & \ldots & L_{1N}(D) \\ L_{21}(D) & L_{22}(D) & \ldots & L_{2N}(D) \\ & \vdots & & \\ L_{N1}(D) & L_{N2}(D) & \ldots & L_{NN}(D) \end{bmatrix} \begin{bmatrix} y_1(t) \\ y_2(t) \\ \vdots \\ y_N(t) \end{bmatrix} = \begin{bmatrix} f_1(t) \\ f_2(t) \\ \vdots \\ f_N(t) \end{bmatrix} \tag{61}$$

where each $L_{ij}(D)$ is a polynomial in $D$, e.g., $L_{12}(D) = (D^3+2D^2+D+5)$. To solve for the variables, we may use Cramer's rule or we may premultiply both sides of (61) by the inverse of the coefficient matrix $\mathbf{L}^{-1}(D)$, to obtain

$$\Delta(D)\begin{bmatrix} y_1(t) \\ y_2(t) \\ \vdots \\ y_N(t) \end{bmatrix} = \Delta(D)\;\mathbf{L}^{-1}(D)\begin{bmatrix} f_1(t) \\ f_2(t) \\ \vdots \\ f_N(t) \end{bmatrix} \tag{62}$$

and we have multiplied both sides by the determinant of $\mathbf{L}(D), \Delta(D)$. From (62) we obtain differential equations in each of the variables $y_1(t), y_2(t), \ldots, y_N(t)$.

*Example 2-5*

Consider the set

$$(D^2+2)y + Dx = 2\cos 2t$$

$$(D+1)y + (D+2)x = 3t$$

Determine differential equations in $y$ and $x$.

*Solution*

In matrix form

$$\begin{bmatrix} (D^2+2) & D \\ (D+1) & (D+2) \end{bmatrix} \begin{bmatrix} y \\ x \end{bmatrix} = \begin{bmatrix} 2\cos 2t \\ 3t \end{bmatrix}$$

The determinant of the coefficient matrix is

$$\Delta(D) = (D^2+2)(D+2) - (D+1)D$$

$$= D^3 + D^2 + D + 4$$

and the inverse is

$$\mathbf{L}^{-1}(D) = \frac{1}{\Delta(D)} \begin{bmatrix} (D+2) & -D \\ -(D+1) & (D^2+2) \end{bmatrix}$$

so that

$$\Delta(D) \begin{bmatrix} y \\ x \end{bmatrix} = \begin{bmatrix} (D+2) & -D \\ -(D+1) & (D^2+2) \end{bmatrix} \begin{bmatrix} 2\cos 2t \\ 3t \end{bmatrix}$$

$$= \begin{bmatrix} -4\sin 2t + 4\cos 2t - 3 \\ 4\sin 2t - 2\cos 2t + 6 \end{bmatrix}$$

with the result

$$(D^3 + D^2 + D + 4)y = -4\sin 2t + 4\cos 2t - 3$$

$$(D^3 + D^2 + D + 4)x = 4\sin 2t - 2\cos 2t + 6t$$

The undetermined constants in the homogeneous solution for each variable are related. For example, in Example 2-5, there are three undetermined constants in the homogeneous solution for $y$ and three in the homogeneous solution for $x$. However these are interrelated. The number of independent, undetermined constants is equal to the degree of the polynomial $\Delta(D)$. Thus the number of independently specifiable initial conditions on $y_1, y_2, \ldots, y_N$ is equal to the degree of $\Delta(D)$.

*Example 2-6*

Consider the set

$$\begin{bmatrix} (D+1) & 1 \\ -2 & (D+4) \end{bmatrix} \begin{bmatrix} y \\ x \end{bmatrix} = \begin{bmatrix} 1 \\ 0 \end{bmatrix}$$

*Solution*

Solving for $y$ and $x$ we obtain

$$(D^2 + 5D + 6)\begin{bmatrix} y \\ x \end{bmatrix} = \begin{bmatrix} (D+4) & -1 \\ 2 & (D+1) \end{bmatrix}\begin{bmatrix} 1 \\ 0 \end{bmatrix} = \begin{bmatrix} 4 \\ 2 \end{bmatrix}$$

or

$$(D^2 + 5D + 6)y = 4$$

$$(D^2 + 5D + 6)x = 2$$

from which we obtain

$$y = C_1 e^{-3t} + C_2 e^{-2t} + \frac{2}{3}$$

$$x = C_3 e^{-3t} + C_4 e^{-2t} + \frac{1}{3}$$

Substitute the result for $y$ into the first equation to yield

$$(D+1)\,[C_1 e^{-3t} + C_2 e^{-2t} + \frac{2}{3}] + x = 1$$

giving

$$x(t) = 2C_1 e^{-3t} + C_2 e^{-2t} + \frac{1}{3}$$

from which we obtain, by comparing to the general solution for $x$,

$$C_3 = 2C_1$$

$$C_4 = C_2$$

and only two initial conditions are independently specifiable. For example, we may specify y(0), $x^{(1)}|_{t_0}$ or $y^{(2)}|_{t_0}$, $x(t_0)$ or y(0), $y^{(1)}|t_0$, etc.

## Solved Problems

*2.1* Consider the electric circuit shown in Fig. P2.1. Determine the current $i(t)$ for $t > 0$ if the initial inductor current is 1A and the initial capacitor voltage is 2V.

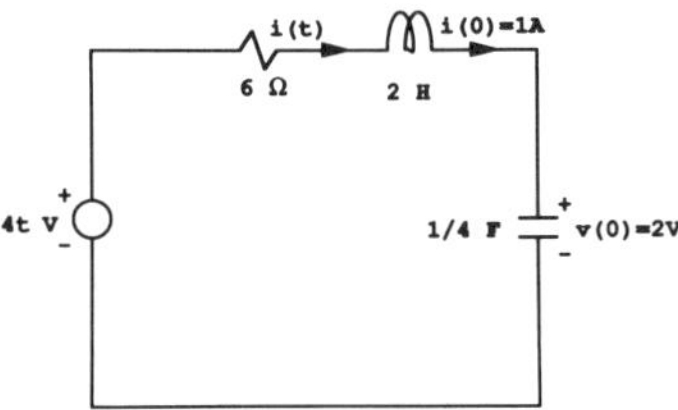


Figure P2.1

*Solution*

Using Kirchoff's voltage law (KVL) around the loop we obtain

$$2\frac{di(t)}{dt} + 6i(t) + 4\int_{-\infty}^{t} i(\tau)\,d\tau = 4t$$

Differentiating we obtain

$$\frac{d^2i(t)}{dt^2} + 3\,\frac{di(t)}{dt} + 2\,i(t) = 2$$

The characteristic equation is

$$p^2 + 3p + 2 = 0$$

so that $p_1 = -2$ and $p_2 = -1$ and

$$i_h(t) = C_1\,e^{-2t} + C_2\,e^{-t}$$

The particular solution is obtained by assuming $i_p = K$. Substituting we obtain $i_p = 2$. Thus

$$i(t) = C_1\,e^{-2t} + C_2\,e^{-t} + 2$$

Since

$$i(0) = 1 = C_1 + C_2 + 2$$

we have

$$C_1 + C_2 = -1$$

The inductor voltage is

$$v(t) = 2\,\frac{di(t)}{dt}$$

Therefore

$$\frac{di}{dt}\Big|_{t=0} = \frac{v(0)}{2}$$

But

$$v(0) = 4t \,|_{t=0} - 6i(0) - 2$$

$$= -12$$

So

$$\frac{di}{dt}\Big|_{t=0} = -2C_1 - C_2$$

$$= -6$$

or

$$2C_1 + C_2 = 6$$

Combining the two equations

$$\begin{bmatrix} 1 & 1 \\ 2 & 1 \end{bmatrix} \begin{bmatrix} C_1 \\ C_2 \end{bmatrix} = \begin{bmatrix} -1 \\ 6 \end{bmatrix}$$

which yields, by any of the methods of Chapter 1,

$$C_1 = 7$$

$$C_2 = -8$$

So

$$i(t) = 7\,e^{-2t} - 8e^{-t} + 2$$

2.2 Consider the circuit in Fig. P2.2. Determine $v(t)$ for $t > 0$ if the initial inductor current is 2A.

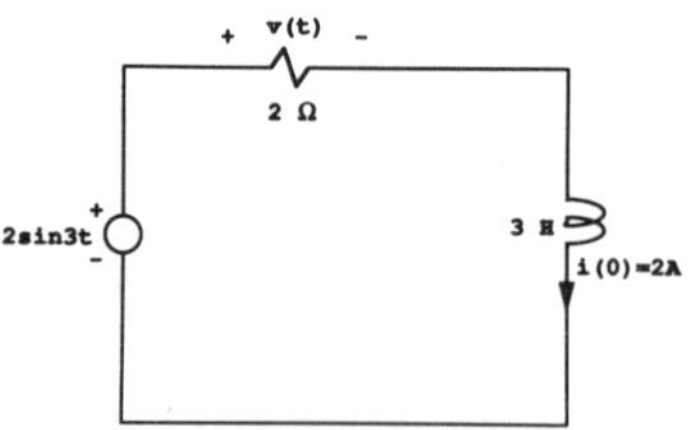


Figure P2.2

*Solution*

With KVL

$$3\frac{di}{dt} + 2\,i = 2\sin 3t$$

or

$$\frac{di}{dt} + \frac{2}{3}\, i = \frac{2}{3} \sin 3t$$

and

$$i_h(t) = C_1\, e^{-2/3t}$$

Assume

$$i_p(t) = K_c \cos 3t + K_s \sin 3t$$

Substitute

$$-9K_c \sin 3t + 9K_s \cos 3t + 2K_c \cos 3t + 2K_s \sin 3t = -2 \sin 3t$$

Matching coefficients

$$-9K_c + 2K_s = 2$$

$$2K_c + 9\, K_s = 0$$

or

$$\begin{bmatrix} -9 & 2 \\ 2 & 9 \end{bmatrix} \begin{bmatrix} K_c \\ K_s \end{bmatrix} = \begin{bmatrix} 2 \\ 0 \end{bmatrix}$$

Solving, for example, by matrix inverse

$$\begin{bmatrix} K_c \\ K_s \end{bmatrix} = -\frac{1}{85} \begin{bmatrix} 9 & -2 \\ -2 & -9 \end{bmatrix} \begin{bmatrix} 2 \\ 0 \end{bmatrix}$$

$$= \begin{bmatrix} -18/85 \\ 4/85 \end{bmatrix}$$

Thus

$$i(t) = C_1\, e^{-2/3t} - \frac{18}{85} \cos 3t + \frac{4}{85} \sin 3t$$

Since

$$i(0) = 2 = C_1 - \frac{18}{85}$$

we find

$$C_1 = \frac{188}{85}$$

and

$$i(t) = \frac{188}{85} e^{-\frac{2}{3}t} - \frac{18}{85} \cos 3t + \frac{4}{85} \sin 3t$$

Since v(t) = 2i(t) we obtain

$$v(t) = \frac{376}{85} e^{-\frac{2}{3}t} - \frac{36}{85} \cos 3t + \frac{8}{85} \sin 3t$$

We could alternatively solve for the particular solution by the method of Section 2.8. Assume

$$\hat{i}_p = K\, e^{j3t}$$

and

$$(j9 + 2) K = 2$$

So

$$K = \frac{2}{2 + j9} = .22\angle -77.47^o$$

and

$$i_p = \text{Im}\,(K\, e^{j3t})$$

$$= .22 \sin (3t - 77.47^o)$$

which is equivalent to

$$i_p = -\frac{18}{85} \cos 3t + \frac{4}{85} \sin 3t$$

*2.3* Determine the capacitor current $i(t)$ in Fig. P2.3 if the initial capacitor voltage is 3V.

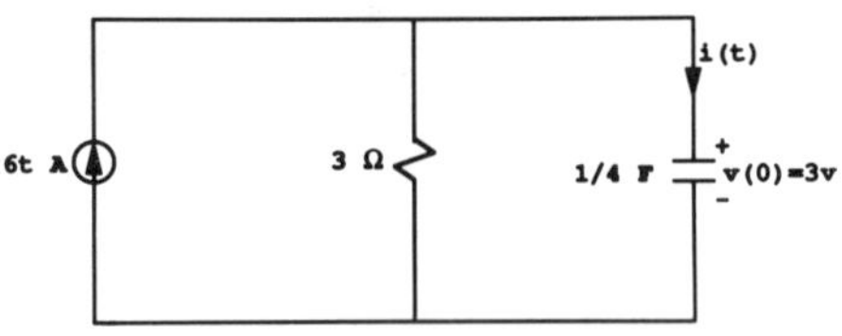


Figure P2.3

*Solution*

Using Kirchoff's current law (KCL) we obtain

$$\tfrac{1}{4}\,\frac{dv}{dt} + \frac{1}{3}\,v = 6t$$

or

$$\frac{dv}{dt} + \frac{4}{3}\,v = 24t$$

The homogeneous solution is

$$v_h(t) = C_1\,e^{-\frac{4}{3}t}$$

The particular solution is obtained by assuming

$$v_p(t) = K_0 + K_1\,t$$

Substituting we obtain

$$K_1 + \frac{4}{3}\,K_0 + \frac{4}{3}\,K_1\,t = 24\,t$$

or

$$K_1 = 18$$

$$K_0 = -\frac{3}{4}\,K_1$$

$$= -\frac{27}{2}$$

so

$$v(t) = C_1\,e^{-\frac{4}{3}t} - \frac{27}{2} + 18t$$

Since $v(0) = 3$

$$v(0) = 3 = C_1 - \frac{27}{2}$$

so

$$C_1 = \frac{33}{2}$$

and

$$v(t) = \frac{33}{2}\,e^{-\frac{4}{3}t} - \frac{27}{2} + 18t$$

Now

$$i(t) = {}^1\!/\!_4 \frac{dv}{dt}$$

$$= -\frac{11}{2} e^{-4/3t} + 9/2$$

*2.4* For the circuit in Fig. P2.4 determine $i(t)$) if the initial inductor current is 2A and the initial capacitor voltage is 3V.

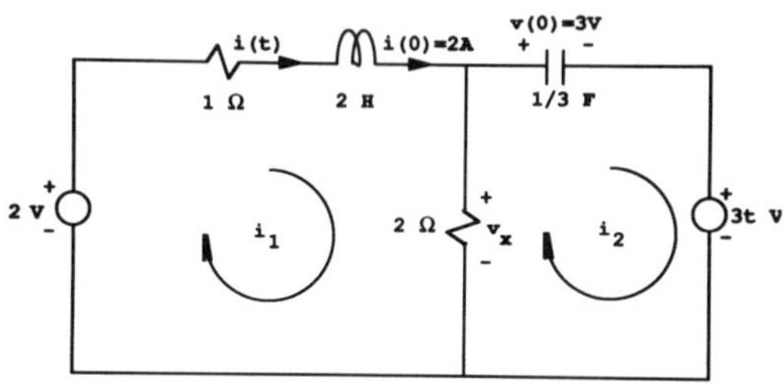


Figure P2.4

*Solution*

Defining mesh currents as shown we obtain

mesh 1: $3i_1 + 2\frac{di_1}{dt} - 2i_2 = 2$

mesh 2: $3 \int_{-\infty}^{t} i_2(\tau)d\tau + 2i_2 - 2i_1 = -3t$

Differentiating the second mesh equation and using the opertor $D$ we obtain

$$(3 + 2D)\, i_1 - 2\, i_2 = 2$$

$$-2D\; i_1 + (2D + 3)\, i_2 = -3$$

In matrix form

$$\begin{bmatrix} (3 + 2D) & -2 \\ -2D & (2D + 3) \end{bmatrix} \begin{bmatrix} i_1 \\ i_2 \end{bmatrix} = \begin{bmatrix} 2 \\ -3 \end{bmatrix}$$

Solving by the method of Section 2.9 we obtain

$$\begin{bmatrix} i_1 \\ i_2 \end{bmatrix} = \frac{1}{4D^2 + 8D + 9} \begin{bmatrix} (2D + 3) & 2 \\ 2D & (3 + 2D) \end{bmatrix} \begin{bmatrix} 2 \\ -3 \end{bmatrix}$$

$$i_1 = \frac{(2D + 3)(2) + (2)(-3)}{4D^2 + 8D + 9}$$

$$i_2 = \frac{(2D)(2) + (3 + 2D)(-3)}{4D^2 + 8D + 9}$$

or

$$\left[D^2 + 2D + \frac{9}{4}\right] i_1 = 0$$

$$\left[D^2 + 2D + \frac{9}{4}\right] i_2 = -9/4$$

The roots of the characteristic equations are $p_1 = -1 - j\sqrt{5/2}$, $p_2 = -1 + j\sqrt{5/2}$.
Thus

$$i_1(t) = e^{-t}\left[C_{11}\cos\frac{\sqrt{5}}{2}t + C_{12}\sin\frac{\sqrt{5}}{2}t\right]$$

$$i_2(t) = e^{-t}\left[C_{21}\cos\frac{\sqrt{5}}{2}t + C_{22}\sin\frac{\sqrt{5}}{2}t\right] - 1$$

Only two of the undetermined constants are independent. Substituting the solution for $i_1$ into the first equation:

$$i_2 = -\tfrac{1}{2}(2) + (3 + 2D)\, i_1$$

$$= -1 + 3\, i_1 + (2D)\, i_1$$

$$= -1 + 3e^{-t}\left[C_{11}\cos\frac{\sqrt{5}}{2}t + C_{12}\sin\frac{\sqrt{5}}{2}t\right]$$

$$-2e^{-t}\left[C_{11}\cos\frac{\sqrt{5}}{2}t + C_{12}\sin\frac{\sqrt{5}}{2}t\right]$$

$$= \sqrt{5}\, e^{-t}\left[-C_{11}\sin\frac{\sqrt{5}}{2}t + C_{12}\cos\frac{\sqrt{5}}{2}t\right]$$

$$= e^{-t}\{(3C_{11} - 2C_{11} + \sqrt{5}\, C_{12})\cos\frac{\sqrt{5}}{2}t$$

$$+ (3C_{12} - 2C_{12} - \sqrt{5}\, C_{11})\sin\frac{\sqrt{5}}{2}t\} - 1$$

Thus we identify

$$C_{21} = C_{11} + \sqrt{5}\, C_{12}$$

$$C_{22} = C_{12} - \sqrt{5}\, C_{11}$$

So we only need to find $C_{11}$ and $C_{12}$ in the solution for $i_1$. We know $i_1(0) = 2A$ and we need $\frac{di_1}{dt}\Big|_{t=0}$. The voltage across the $2\Omega$ resistor, $v_x$ is, by KVL around the second loop,

$$v_x(0) = 3 + 3t \,\Big|_{t=0} = 0$$

Summing KVL around the first loop

$$2\frac{di_1}{dt} + i_1 = 2 - v_x$$

and

$$\frac{di_1}{dt}\Big|_{t=0} = -\tfrac{1}{2}\, i_1(0) + 1 - \tfrac{1}{2}\, v_x(0)$$

$$= -3/2$$

Thus

$$i_1(0) = C_{11} = 2$$

$$\frac{di_1}{dt} = -e^{-t}\left(C_{11}\cos\frac{\sqrt{5}}{2}t + C_{12}\sin\frac{\sqrt{5}}{2}t\right)$$

$$+ \frac{\sqrt{5}}{2}e^{-t}\left(-C_{11}\sin\frac{\sqrt{5}}{2}t + C_{12}\cos\frac{\sqrt{5}}{2}t\right)$$

and

$$\frac{di_1}{dt}\Big|_{t=0} = -C_{11} + \frac{\sqrt{5}}{2}C_{12}$$

$$= -3/2$$

so

$$C_{12} = \frac{2}{\sqrt{5}}C_{11} - \frac{3}{\sqrt{5}}$$

$$= \frac{1}{\sqrt{5}}$$

Thus

$$i_1(t) = e^{-t}\left(2\cos\frac{\sqrt{5}}{2}t + \frac{1}{\sqrt{5}}\sin\frac{\sqrt{5}}{2}t\right)$$

2.5 Consider the spring, mass, dashpot system in Fig. P2.5. Determine the equation for the mass position if the initial position is $x(0) = 2$, the initial velocity $v(0) = 1$ and the applied force is $2t$.

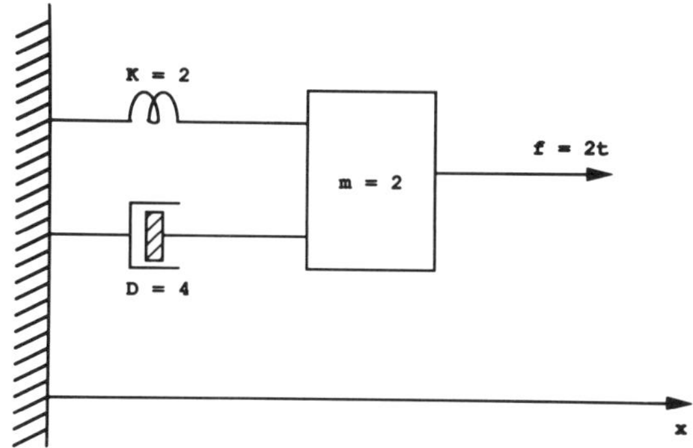


Figure P2.5

*Solution*

Summing the forces applied to the mass yields, by D'Alembert's principle,

$$m \frac{d^2x}{dt^2} + D \frac{dx}{dt} + K x = f$$

or

$$\frac{d^2x}{dt^2} + 2 \frac{dx}{dt} + x = t$$

The characteristic equation is

$$p^2 + 2p + 1 = 0$$

The roots are

$$p_1, p_2 = -1$$

Thus

$$x_h = C_1 e^{-t} + C_2 te^{-t}$$

The particular solution is found by assuming $x_p(t) = K_0 + K_1 t$ so that

$$2K_1 + K_0 + K_1 t = t$$

gives

$$K_1 = 1$$

$$K_0 = -2$$

and

$$x(t) = C_1 e^{-t} + C_2 t e^{-t} - 2 + t$$

Applying the initial condition, $x(0) = 2$, yields

$$x(0) = 2 = C_1 - 2$$

or $C_1 = 4$. Similarly

$$v(t) = \frac{dx}{dt} = -C_1 e^{-t} - C_2 t e^{-t} + C_2 e^{-t} + 1$$

and

$$v(0) = 1 = -C_1 + C_2 + 1$$

or $C_2 = C_1 = 4$. Thus

$$x(t) = 4e^{-t}(1 + t) - 2 + t$$

*2.6* A small mass is supported by vertical wires as shown in Fig. P2.6. The mass is displaced horizontally a small distance and released. Determine the frequency of vibration of the mass.

*Solution*

The mass is displaced a small distance $x$ and is acted upon by the tensions $T$. For small displacements the $x$ components will be $T\frac{x}{b}$ and $T\frac{x}{c}$ both to the left. Thus

$$-T\frac{x}{b} - T\frac{x}{c} = M\frac{d^2x}{dt^2}$$

or

$$\frac{d^2x}{dt^2} + \frac{T}{M}\left(\frac{1}{b} + \frac{1}{c}\right)x = 0$$

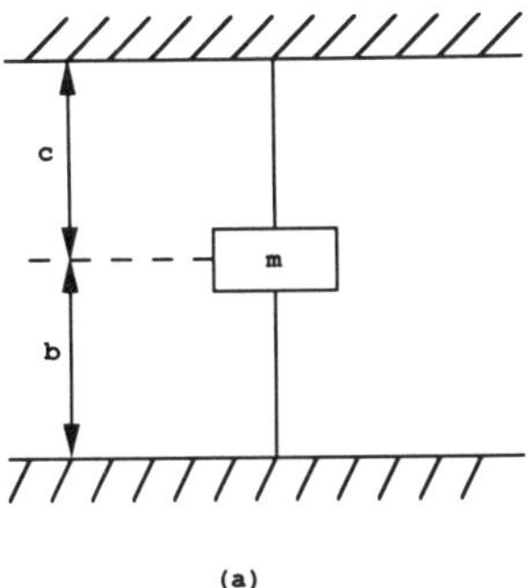


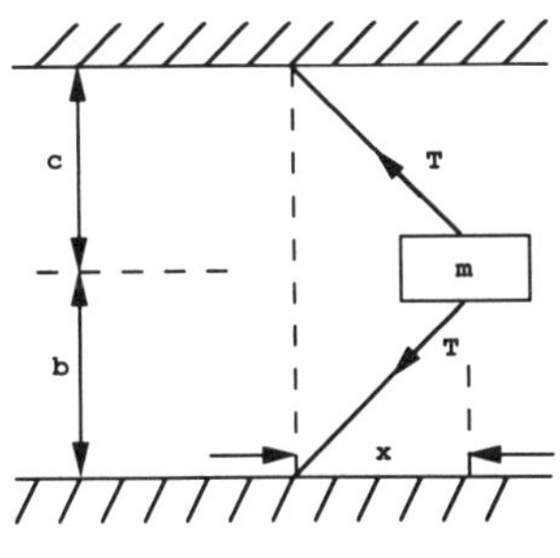


Figure P2.6

Only a homogeneous solution will exist and the characteristic equation becomes

$$p^2 + \frac{T}{M}\left(\frac{1}{b} + \frac{1}{c}\right) = 0$$

with roots

$$p_1, p_2 = \pm\, j\sqrt{\frac{T}{M}\left(\frac{1}{b} + \frac{1}{c}\right)}$$

Thus

$$x_h(t) = C_1 \cos\sqrt{\frac{T}{M}\left(\frac{1}{b} + \frac{1}{c}\right)}\, t + C_2 \sin\sqrt{\frac{T}{M}\left(\frac{1}{b} + \frac{1}{c}\right)}\, t$$

and we identify the natural frequency of oscillation as

$$f = \frac{1}{2\pi}\sqrt{\frac{T}{M}\left(\frac{1}{b}+\frac{1}{c}\right)} \quad Hz$$

2.7 A disk having a moment of inertia $I_0$ is rigidly attached to a slender shaft of torsional stiffness $K$ as shown in Fig. P2.7, where $K$ is the torque necessary to twist the shaft through one radian. Determine the natural frequency of oscillation if the shaft is twisted through a small angle and then released.

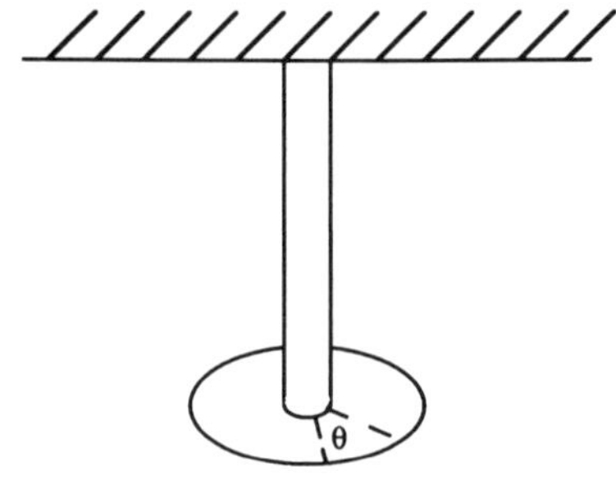


Figure P2.7

*Solution*

When the disc is twisted through a small angle $\theta$, there is a restoring force $K\theta$ exerted which is balanced by the restoring force $I_0 \dfrac{d^2\theta}{dt^2}$. Thus

$$\frac{d^2\theta}{dt^2} + \frac{K}{I_0}\theta = 0$$

Thus only a homogeneous solution exists, and the roots of the characteristic equation are

$$p_1, p_2 = \pm j\sqrt{\frac{K}{I_0}}$$

Thus

$$\theta_h(t) = C_1 \cos\sqrt{\frac{K}{I_0}}\,t + C_2 \sin\sqrt{\frac{K}{I_0}}\,t$$

and the natural frequency of oscillation is identified as

$$f = \frac{1}{2\pi}\sqrt{\frac{K}{I_0}} \quad Hz$$

*2.8* A mass $m$ is pivoted on an arm of length $\ell$ as shown in Fig. P2.8(a). A spring and a dashpot are attached as shown. If the mass is pulled downward a small amount as shown in Fig. P.2.8(b) and released, determine the equation of motion in terms of angle θ and the natural frequency of oscillation.

*Solution*

The free body diagrams are shown in Fig. P.2.8(b). Summing moments about 0 and equating to $m\ell^2 \dfrac{d^2\theta}{dt^2}$

gives

$$mg\ell - Kb^2(\theta_0 + \theta) - Db^2 \frac{d\theta}{dt} = m\ell^2 \frac{d^2\theta}{dt^2}$$

But $mg\ell = Kb^2\theta_0$ at equilibrium. Thus the equation of motion for small oscillation is

$$\frac{d^2\theta}{dt^2} + \frac{Db^2}{m\ell^2}\frac{d\theta}{dt} + \frac{Kb^2}{m\ell^2}\theta = 0$$

Substituting the values $K = 9, D = 8, m = 2, b = 1, \ell = 2$ we have

$$\frac{d^2\theta}{dt^2} + 2\frac{d\theta}{dt} + 2\theta = 0$$

The characteristic equation

$$p^2 + 2p + 2 = 0$$

has roots

$$p_1, p_2 = -1 \pm j\,1$$

so that

$$\theta_h(t) = e^{-t}\,(C_1 \cos t + C_2 \sin t)$$

Thus the frequency of oscillation is

$$f = \frac{1}{2\pi} \quad Hz$$

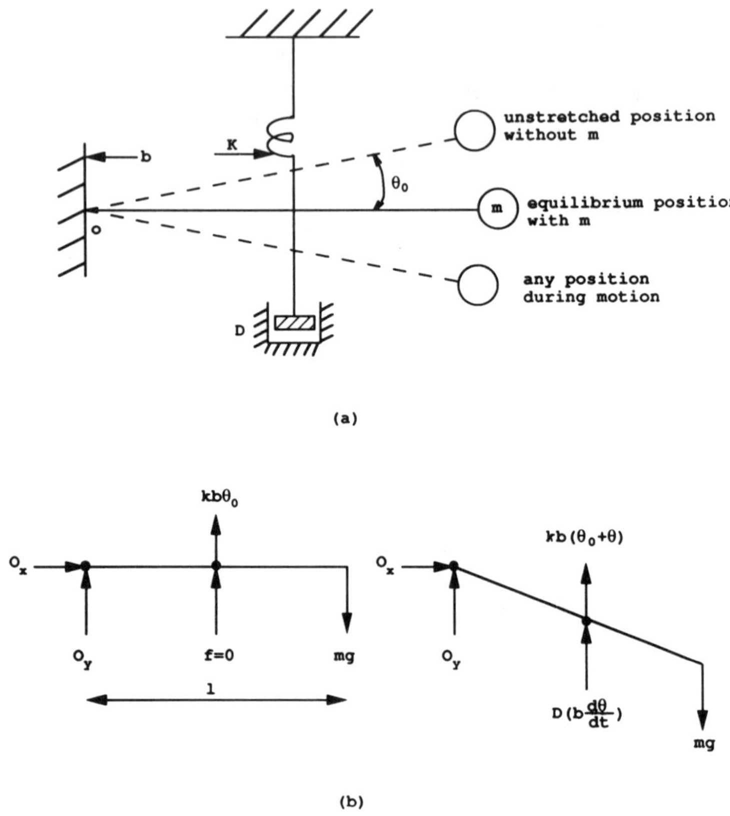


Figure P2.8

*2.9* Determine $x_1(t)$ and $x_2(t)$ which describe the motion of the mechanical system in Fig. P2.9(a). The undetermined constants in the homogeneous solution will not be determined.

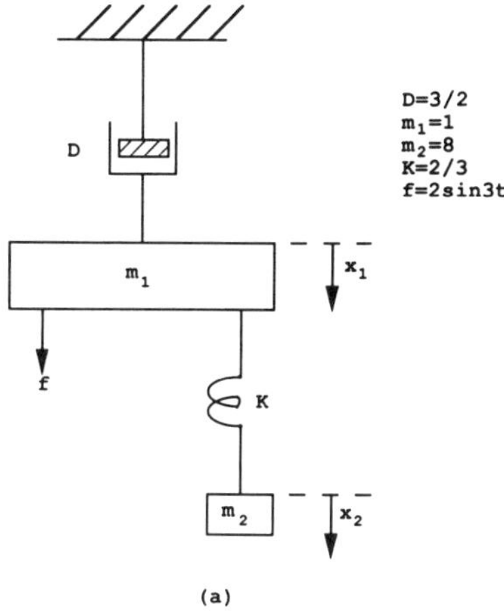


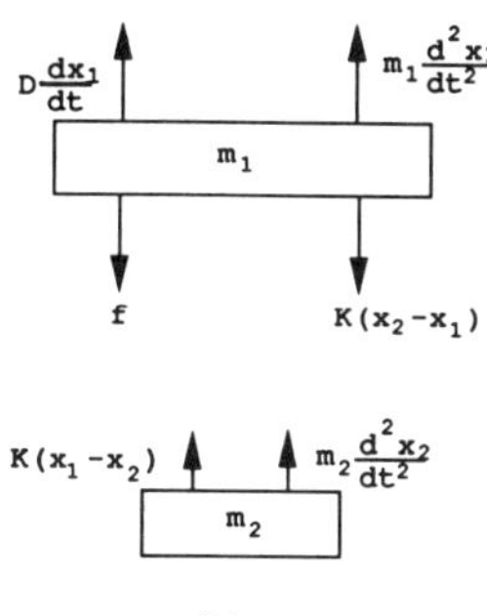


Figure P2.9

*Solution*

The free body diagrams are shown in Fig. P2.9(b). Applying D'Alembert's principle we obtain

$$m_1 \frac{d^2x_1}{dt^2} + D \frac{dx_1}{dt} + K\, x_1 - K\, x_2 = f$$

$$-K\, x_1 + m_2 \frac{d^2x_2}{dt^2} + K\, x_2 = 0$$

With the operator $D$ and substituting values we have

$$\begin{bmatrix} (D^2 + \frac{3}{2}D + \frac{2}{3}) & -2/3 \\ -2/3 & (8D^2 + 2/3) \end{bmatrix} \begin{bmatrix} x_1 \\ x_2 \end{bmatrix} = \begin{bmatrix} 2\sin 3t \\ 0 \end{bmatrix}$$

Solving, for example, by Cramer's rule

$$\begin{vmatrix} (D^2 + 3/2\,D + 2/3) & -2/3 \\ -2/3 & (8D^2 + 2/3) \end{vmatrix} = 8D^4 + 12D^3 + \frac{18}{3}D^2 + D$$

$$x_1 = \frac{\begin{vmatrix} 2\sin 3t & -2/3 \\ 0 & (8D^2 + 2/3) \end{vmatrix}}{8D^4 + 12D^3 + \frac{18}{3}D^2 + D}$$

$$= \frac{-144\sin 3t + \frac{4}{3}\sin 3t}{8D^4 + 12D^3 + \frac{18}{3}D^2 + D}$$

$$x_2 = \frac{\begin{vmatrix} (D^2 + 3/2D + 2/3) & 2\sin 3t \\ -2/3 & 0 \end{vmatrix}}{8D^4 + 12D^3 + \frac{18}{3}D^2 + D}$$

$$= \frac{4/3\sin 3t}{8D^4 + 12D^3 + \frac{18}{3}D^2 + D}$$

Thus

$$(D^4 + \frac{3}{2}D^3 + \frac{3}{4}D^2 + \frac{1}{8}D)\,x_1 = -18\sin 3t + \frac{1}{6}\sin 3t$$

$$= -\frac{107}{6}\sin 3t$$

$$(D^4 + \frac{3}{2}D^3 + \frac{3}{4}D^2 + \frac{1}{8}D)\,x_2 = \frac{1}{6}\sin 3t$$

The characteristic equation

$$p^4 + \frac{3}{2}p^3 + \frac{3}{4}p^2 + \frac{1}{8}p = 0$$

has roots

$$p_1 = 0,\ p_2 = p_3 = p_4 = -1/2$$

(High order characteristic equations for physical systems rarely factor this nicely. Practically speaking, a digital computer is needed to factor the characteristic equation.) Thus

$$x_{1h}(t) = C_{11} + C_{12}\,e^{-t/2} + C_{13}\,te^{-t/2} + C_{14}\,t^2e^{-t/2}$$

$$x_{2h}(t) = C_{21} + C_{22}e^{-t/2} + C_{23}te^{-t/2} + C_{24}\, t^2\, e^{-t/2}$$

The particular solutions are found from the method of Section 2.8 by assuming

$$\hat{x}_{1p} = K_1\, e^{j3t}$$

$$\hat{x}_{2p} = K_2\, e^{j3t}$$

Thus

$$K_1 = \frac{-107/6}{(j3)^4 + \frac{3}{2}(j3)^3 + \frac{3}{4}(j3)^2 + \frac{1}{8}(j3)} = -.21\angle 28.4^o$$

$$K_2 = \frac{1/6}{(j3)^4 + \frac{3}{2}(j3)^3 + \frac{3}{4}(j3)^2 + \frac{1}{8}(j3)} = 1.975 \times 10^{-3}\angle 28.4^o$$

Thus

$$x_{1p} = 0.21 \sin(3t + 28.4^o)$$

$$x_{2p} = 1.975 \times 10^{-3} \sin(3t + 28.4^o)$$

The total solution is

$$x_1(t) = C_{11} + C_{12}\, e^{-t/2} + C_{13}\, te^{-t/2} + C_{14}\, t^2\, e^{-t/2} - .21 \sin(3t + 28.4^o)$$

$$x_2(t) = C_{21} + C_{22}\, e^{-t/2} + C_{23}\, te^{-t/2} + C_{24}\, t^2 e^{-t/2} +$$

$$1.975 \times 10^{-3} \sin(3t + 28.4^o)$$

Since the order of the characteristic equation is four, only four of these undetermined constants are independent. In order to determine the relationships amoung the undetermined constants, substitute $x_2(t)$ into the second equation:

$$-\frac{2}{3}\, x_1(t) + (8D^2 + 2/3)\, x_2(t) = 0$$

or

$$x_1(t) = (12D^2 + 1)\, x_2(t)$$

$$= C_{21} + (C_{22} - 12\, C_{23} + 24C_{24})\, e^{-t/2}$$

$$+ (4C_{23} - 24\, C_{24})\, te^{-t/2}$$

$$+ (4C_{24})\, t^2 e^{-t/2} - 0.21 \sin(3t + 28.4^o)$$

Thus we identify

$$C_{11} = C_{21}$$

$$C_{12} = C_{22} - 12\,C_{23} + 24\,C_{24}$$

$$C_{13} = 4C_{23} - 24C_{24}$$

$$C_{14} = 4C_{24}$$

Therefore we may specify

$$x_1(0),\ v_1(0),\ v_1^{(1)}(0),\ v_1^{(2)}(0)$$

or

$$x_1(0),\ x_2(0),\ v_1(0),\ v_2(0)$$

etc., to evaluate the undetermined constants in the homogeneous solutions.

*2.10* A body initially at temperature $T_i$ is suddenly immersed in a fluid of infinite extent which is at temperature $T_\infty$. Determine and solve the differential equation describing the temperature of the body.

*Solution*

The heat $Q$ is related to the heat transfer coefficient h by

$$Q(t) = h\,A\,(T(t) - T_\infty) \tag{a}$$

where A is the surface area of the body. Also

$$Q(t) = -\frac{d}{dt}\,(\textit{internal energy})$$

$$= -\frac{d}{dt}\,(\rho\,V\,c\,T(t)) \tag{b}$$

where $\rho$ is the body density, $V$ is the volume, and $c$ is the specific heat. Equating (a) and (b) we obtain

$$\frac{dT(t)}{dt} = -\frac{hA}{\rho Vc}\,(T(t) - T_\infty)$$

Defining

$$M = \frac{hA}{\rho Vc}$$

and

$$\theta(t) = \frac{T(t) - T_\infty}{T_i - T_\infty}$$

we obtain

$$\frac{d\theta(t)}{dt} + M\ \theta(t) = 0$$

Thus only a homogeneous solution exists:

$$\theta(t) = C\ e^{-Mt}$$

Note that

$$\theta(0) = \frac{T(0) - T_\infty}{T_i - T_\infty}$$

$$= 1$$

Thus $C = 1$ and

$$\theta(t) = e^{-Mt}$$

*2.11* A wide thin fin as shown in Fig. P2.11(a) transfers heat by convection. Assume a uniform heat transfer coefficient $h$ to the medium which is at temperature $T_\infty$. The base of the fin is at temperature $T_B$. Determine the temperature distribution over the cross section of the fin.

*Solution*

Considering the energy balance for a $\Delta x$ section of the fin as shown in Fig. P2.11(b) we obtain

$$Q^{(x)}_{cond_x} - Q^{(x)}_{cond_{x+\Delta x}} = h\,\Delta x\ P\ \ (T(x) - T_\infty)$$

where $P$ is the perimeter of this $\Delta x$ slice of the fin given by $P = 2\,(W + \Delta\,x)$. Dividing both sides by $\Delta x$ and taking the limit as $\Delta x \to 0$ yields

$$\frac{d\,Q\,(x)}{dx} = -hP\ \ (T(x) - T_\infty)$$

From Fourier's law

$$Q\,(x) = -kA\ \frac{dT(x)}{dx}$$

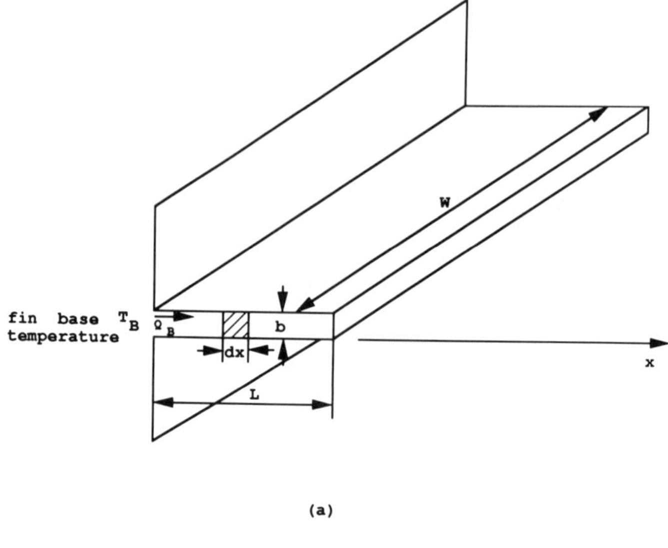


(a)

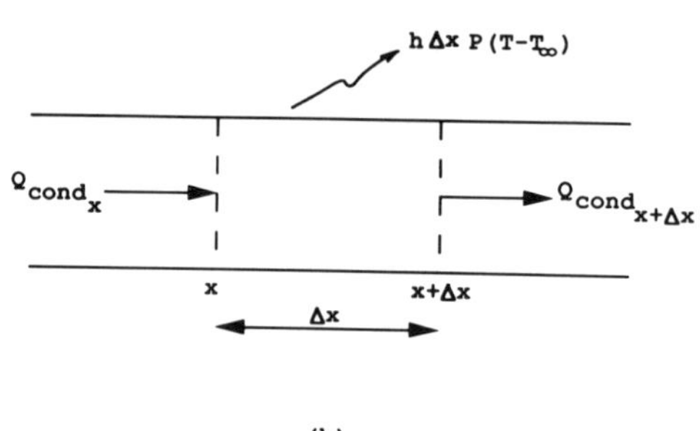


(b)

Figure P2.11

where $k$ is the thermal conductivity of the fin and $A$ is $bW$. Thus

$$\frac{d^2T(x)}{dx^2} = \frac{hP}{kA}\,(T(x) - T_\infty)$$

Defining

$$\theta(x) = \frac{T(x) - T_\infty}{T_B - T_\infty}$$

we obtain

$$\frac{d^2\theta(x)}{dx^2} - M\,\theta(x) = 0$$

where

$$M = \frac{hP}{kA}$$

The characteristic equation is

$$p^2 - M = 0$$

with roots $p_1 = \sqrt{M}$, $p_2 = -\sqrt{M}$. Thus the homogeneous solution is the total solution and

$$\theta(x) = C_1 e^{\sqrt{M}\,x} + C_2 e^{-\sqrt{M}\,x}$$

In order to determine the constants $C_1$ and $C_2$ we need two conditions. The first is that $T(0) = T_B$. The second condition is usually one of the following:

(1) a long, thin fin with $T(L) = T_\infty$,

(2) a fin not quite so long with $Q(L) = 0$ such that

$$\frac{dT(x)}{dx}\Big|_{x=L} = 0$$

(3) a $x = L$ the end loses heat with a heat transfer coefficient $h_e$. From $T(0) = T_B$ we obtain

$$\theta(0) = 1$$

$$= C_1 + C_2$$

Using condition (1) gives

$$\theta(L) = 0$$

$$= C_1 e^{\sqrt{M}L} + C_2 e^{-\sqrt{M}L}$$

so that

$$\begin{bmatrix} 1 & 1 \\ \sqrt{ML} & -\sqrt{ML} \\ 3 & 3 \end{bmatrix} \begin{bmatrix} C_1 \\ C_2 \end{bmatrix} = \begin{bmatrix} 1 \\ 0 \end{bmatrix}$$

or

$$C_1 = \frac{e^{-\sqrt{ML}}}{e^{-\sqrt{ML}} - e^{\sqrt{ML}}}$$

$$C_2 = -\frac{e^{\sqrt{ML}}}{e^{-\sqrt{ML}} - e^{\sqrt{ML}}}$$

and

$$\theta(x) = \frac{e^{\sqrt{M}\,(x-L)} - e^{-\sqrt{M}\,(x-L)}}{e^{-\sqrt{M}L} - e^{\sqrt{M}L}}$$

If we had applied $T(0) = T_B$ and condition (2) then

$$\theta(0) = 1 = C_1 + C_2$$

but

$$\frac{dT}{dx}\Big|_{x=L} = 0$$

gives

$$\frac{d\theta}{dx}\Big|_{x=L} = \frac{1}{T_B - T_\infty}\frac{dT}{dx}\Big|_{x=L}$$

$$= 0$$

so that

$$0 = \sqrt{M}\,C_1 - \sqrt{M}\,C_2$$

and

$$\begin{bmatrix} 1 & 1 \\ \sqrt{M} & -\sqrt{M} \end{bmatrix} \begin{bmatrix} C_1 \\ C_2 \end{bmatrix} = \begin{bmatrix} 1 \\ 0 \end{bmatrix}$$

Thus

$$C_1 = \frac{-\sqrt{M}}{-2\sqrt{M}} = 1/2$$

$$C_2 = \frac{-\sqrt{M}}{-2\sqrt{M}} = 1/2$$

and

$$\theta(x) = {}^1\!/_2\, e^{\sqrt{M}x} + {}^1\!/_2\, e^{-\sqrt{M}x}$$

$$= \cos(\sqrt{M}\,x)$$

In either case the heat at the boundary $Q_B$ is

$$Q_B = -kA\,\frac{dT}{dx}\Big|_{x=0}$$

$$= -kA\,(T_B - T_\infty)\,\frac{d\theta}{dx}\Big|_{x=0}$$

CHAPTER

# 3

# PARTIAL DIFFERENTIAL EQUATIONS

## 3.1 Introduction

Whereas ordinary differential equations are applicable to lumped-parameter systems problems (discussed in Chapter 2) partial differential equations arise in physical problems involving distributed parameters. For instance, the dynamics of an electric circuit, having the parameters $R$, $L$, and $C$ (the resistance, inductance and capacitance respectively), is governed by an ordinary differential equation with constant coefficients. Similarly, the mechanical motion of a spring, mass and dashpot system is given by an ordinary differential equation. On the other hand, the phenomenon of heat flow in a solid, the propagation of electromagnetic waves in space, and the longitudinal vibration of a bar are distributed-parameter phenomena and are properly described by the solutions to the pertinent partial differential equation. In this chapter, we shall present a wide variety of partial differential equations governing diverse physical phenomena in a host of engineering problems. Although we shall derive a few of these equations, the emphasis will be on obtaining solutions to these equations. Theorems concerning uniqueness and existence of solutions (and their uniform convergence) are not included here.

## 3.2 Basic Concepts

Partial differential equations involve partial derivatives just as ordinary differential equations involve differential coefficients or total derivatives. Thus, a partial differential equation is an equation involving one or more partial derivatives of a function of two or more independent variables. In engineering problems, the variables are the space variables: $(x,y,z)$, $(r,\theta,z)$ and $(r,\theta,\phi)$ depending upon the chosen system of coordinates; and time, $t$ (for problems involving time variation, i.e. dynamic problems).

As we have seen for an ordinary differential equation (Chapter 2), a partial differential equation is *linear* if the dependent variable and its partial derivatives occur in the first degree. If each term of the equation contains either the dependent variable or one of its derivatives, the equation is *homogeneous* (otherwise the equation is nonhomogeneous). The order of the highest derivative determines the *order* of the equation.

## 3.3 Solution to an Engineering Problem

Essential steps in obtaining a solution to an engineering problem governed by a partial differential equation are as follows:

1. Express the pertinent phenomenon as a partial differential equation. This is obtained by an application of the governing physical laws.
2. Specify the region(s) where the equation is valid.
3. Specify the boundary conditions.
4. Solve the equation subject to the given boundary conditions.
5. Intrepret the solution.

We now illustrate these steps by means of an example.

*Example 3.1*

Problems relating to potential distributions, such as electrostatic, magnetic, thermal, etc., arise in many branches of engineering. In a two-dimensional rectangular space, of size $a$ by $b$, the electrostatic scalar potential distribution is to be found. The potentials on the four sides of the rectangle are as shown in Fig. 3.1. Proceed as in the steps outlined above.

*Solution*

Step 1. Let $\psi = \psi(x,y)$ be the potential within the given region. Applying the laws of electrostatics yields the following governing equation (see Problem 3.1).

$$\frac{\partial^2 \psi}{\partial x^2} + \frac{\partial^2 \psi}{\partial y^2} = 0 \tag{1}$$

Step 2. This equation (known as Laplace's equation) is valid withing the rectangular region.

Step 3. The boundary conditions are (from Fig. 3.1):

$$\psi(0,y) = 0$$

$$\psi(b,y) = \psi$$

$$\psi(x,0) = 0$$

$$\psi(x,a) = 0$$

Step 4. The complete solution is given by (See Problem 3.3).

$$\psi = \frac{4\psi_0}{\pi} \sum_{n,\,odd} \frac{1}{n} \frac{\sinh \frac{\pi n x}{a}}{\sinh \frac{\pi n b}{a}} \sin \frac{\pi n y}{a} \tag{2}$$

Step 5. Equipotentials are plotted in Fig. 3.1 from (2)

### 3.4 Some Commonly Encountered Equations

Deferring the derivations to the problems at the end of the chapter, the following are some of the commonly encountered partial differential equations and their applications to engineering problems.

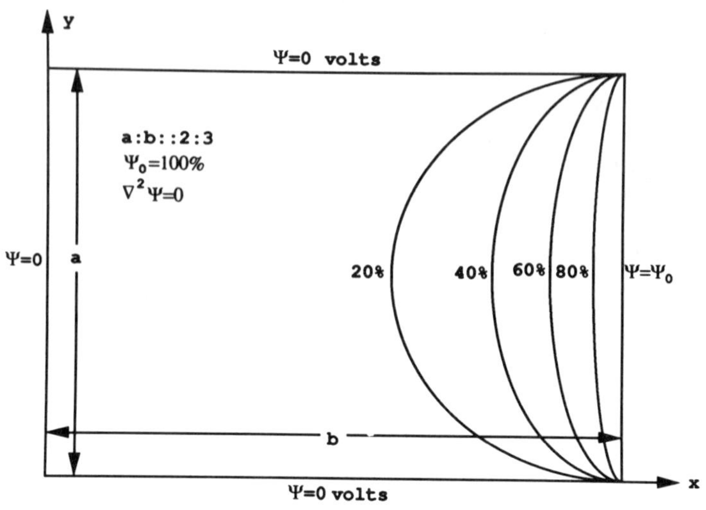


Figure 3.1

| Equation | Some Applications |
|---|---|
| (1) Laplace's equation: $\nabla^2\psi = 0$ $\nabla^2$ being the Laplacian operator. | (a) fluid flow, with no vorticity ($\psi$ = streamlines); (b) electrostatics, in a source-free region ($\psi$ = electric potential). |
| (2) Poisson's equation: $\nabla^2\psi = -\alpha$ | (a) fluid dynamics with $\psi$ = stream function and $\alpha$ = vorticity; (b) electrostatics, with $\psi$ = electric potential and $\alpha$ = ratio of charge density to dielectric constant; (c) elasticity, with $\alpha = 2$, from $\psi$ the torsion of a cylinder can be calculated. |
| (3) Wave equation: $a^2\frac{\partial^2\phi}{\partial x^2} = \frac{\partial^2\phi}{\partial t^2}$ where a = a constant. [Or, $a^2\nabla^2\phi = \frac{\partial^2\phi}{\partial t^2}$ in three dimensions] | (a) vibrating string; (b) electromagnetic waves, with $a$ = velocity of propagation of $\phi$; (c) torsional vibration of a shaft, with $\sqrt{a}$ = ratio of the rigidity modulus of the shaft to its mass density and $\phi$ = angle of twist; (d) wave phenomena in general. |
| (4) Vibrating beam equation: $\frac{\partial^2 y}{\partial t^2} + \frac{EI}{A\rho}\frac{\partial^4 y}{\partial x^4} = 0$ where $E$ = modulus of elasticity, | Transverse vibrations of a beam. |

| | |
|---|---|
| $I$ = cross-sectional moment of inertia,<br>$A$ = cross-sectional area, $\rho$ = density<br>and $y$ = beam deflection. | |
| (5) Diaphragm vibration equation:<br>$\frac{\partial^2 u}{\partial x^2} + \frac{1}{x}\frac{\partial u}{\partial x} = a\frac{\partial^2 u}{\partial t^2} + b\frac{\partial u}{\partial t}$<br>where $u$ = diaphragm displacement<br>and $a$, $b$ = constants. | Vibration of a capacitor microphone diaphragm. |
| (6) Diffusion equation (or heat flow equation):<br>$\nabla^2 \rho = k^2 \frac{\partial \rho}{\partial t}$<br>where $k$ = a constant. | Diffusion of time-varying electric currents in metallic conductors and other diffusion processes, including heat flow and interdiffusion of two substances. |
| (7) Transmission-line equation:<br>$\frac{\partial^2 V}{\partial x^2} = LC\frac{\partial^2 V}{\partial t^2} + (RC + GL)\frac{\partial V}{\partial t} + RGV$<br>where $R$, $L$, $C$ and $G$ are respectively the resistance, inductance, capacitance and conductance per unit length of the line. | Voltage distribution on a non-ideal long electric transmission line. |
| (8) Biharmonic equation:<br>$\nabla^4 y = \frac{q}{D}$<br>where $y$ is the transverse deflection of a thin plate of flexural rigidity $D$ subject to a normal load $q$ per unit area. | Transverse deflection of a thin plate. |
| (9) Vorticity transport equation:<br>$\frac{\partial \zeta}{\partial t} + u\frac{\partial \zeta}{\partial x} + v\frac{\partial \zeta}{\partial y} = \nu \nabla^2 \zeta$<br>where $\nu$ is the kinematic viscosity and $u$ and $v$ are the velocity components in $x$ and $y$ directions. | Vorticity $\zeta$ of an incompressible fluid in two-dimensional motion. |
| (10) Laminar flow equations:<br>$u\frac{\partial u}{\partial x} + v\frac{\partial u}{\partial y} = -\frac{1}{\rho}\frac{\partial p}{\partial x} + \frac{\nu}{\rho}\nabla^2 u$<br>$u\frac{\partial v}{\partial x} + v\frac{\partial v}{\partial y} = -\frac{1}{\rho}\frac{\partial p}{\partial x} + \frac{\nu}{\rho}\nabla^2 v$<br>$\frac{\partial u}{\partial x} + \frac{\partial v}{\partial y} = 0$<br>where $p$ = pressure, $\rho$ = density, $\nu$ = coefficient of viscosity, and $u$, $v$ = velocity components. | Fluid dynamics. |
| (11) Laminar flow heat-exchanger equation:<br>$\frac{\partial^2 T}{\partial r^2} + \frac{1}{r}\frac{\partial T}{\partial r} + \frac{\partial^2 T}{\partial z^2} = \frac{\rho u c_p}{k}\frac{\partial T}{\partial z}$ | Variation of temperature $T(r, z)$ for steady laminar flow in a cylindrical |

| | |
|---|---|
| where $\rho$ = density, $c_p$ = specific heat, $k$ = thermal conductivity, and $u$ = axial velocity. | heat exchanger. |
| (12) Ion-exchange equation: $v\frac{\partial c}{\partial x} + \varepsilon\frac{\partial c}{\partial t} + r = 0$ where $c$ = concentration of a particular ion in solution, $r$ = rate at which that ion is adsorbed per unit volume of a column of ion-exchange resin, $\varepsilon$ = bed void fraction, and $v$ = superficial liquid velocity. | (a) Flow of a solution through a packed column containing ion-exchange resin; (b) Regenerative heat transfer. |

The above list is intended to be a sampler, and is by no means complete. We see from above that the major physical phenomena governed by partial differential equations belong to one of the following categories:

1. Fields - such as electric, gravitational, thermal, and elastic.
2. Mechanical motion - such as vibration, deflection, displacement, and oscillation.
3. Transfer - transport - such as diffusion, heat transfer and fluid flow. We now consider some of the methods of solving these equations.

## 3.5 Methods of Solving Partial Differential Equations

In general, solutions to partial differential equations can be obtained by: direct integration; separation of variables; transform method such as Laplace transform; and numerical methods. The method of direct integration has a very limited application whereas separation of variables is applicable to a large class of problems known as boundary value problems. Transform method can be conveniently used to solve linear partial differential equations (in which the unknown function and its derivative occur in the first degree). Numerical methods are approximate solutions and yield numbers (in contrast to analytical expressions) as end results. Numerical methods are presented in Chapter 5, and in the following we consider analytical solutions to partial differential equations.

### 3.5.1 D'Alembert Solution to the Wave Equation

A general solution to the wave equation (3) can be obtained

$$\frac{\partial^2 u}{\partial t^2} = c^2 \frac{\partial^2 u}{\partial x^2}, \quad c = \text{a constant} \tag{3}$$

by direct integration through a change of variables. So let

$$v = x + ct \quad \text{and } z = x - ct \tag{4}$$

In (3) we have $u = u(x,t)$ whereas in (4), $u = u(v,z)$. Realizing that $\partial v / \partial x = \partial z / \partial x = 1$, (3) and (4), upon applying the chain rule, we obtain

$$\frac{\partial^2 u}{\partial x^2} = \frac{\partial^2 u}{\partial v^2} + 2\frac{\partial^2 u}{\partial v \partial z} + \frac{\partial^2 u}{\partial z^2} \tag{5a}$$

and

$$\frac{\partial^2 u}{\partial t^2} = c^2 \left[\frac{\partial^2 u}{\partial v^2} - 2\frac{\partial^2 u}{\partial v \partial z} + \frac{\partial^2 u}{\partial z^2}\right] \tag{5b}$$

Hence, (3) and (5a,b) yield

$$\frac{\partial^2 u}{\partial v \partial z} = 0 \tag{6}$$

Clearly (6) can be readily integrated. Thus, integrating (6) with respect to $z$ gives

$$\frac{\partial u}{\partial v} = \phi(v) \tag{7}$$

where $\phi(v)$ is an arbitrary function of $v$. Integrating (7) with respect to $v$, we get

$$u = \int \phi(v) dv + g(z) \tag{8}$$

where $g(z)$ is an arbitrary function of $z$. Notice from (7) and (8) that integrating a second order partial differential equation requires two arbitrary functions (rather than two arbitrary constants, as in a second order ordinary differential equation). Now, in (8) we may let $\int \phi(v) dv = f(v)$, a function of $v$ only, the solution becomes

$$u = f(v) + g(z)$$

or,

$$u(x,t) = f(x + ct) + g(x - ct) \tag{9}$$

which is known as d'Alembert solution to (3).

### 3.5.2 Separation of Variables

The method of separation of variables is a general method applicable to homogeneous linear, partial differential equation, including (3) solved above (See Problems 3.2 and 3.3). The method implies that a solution of the form $\psi = \psi\ (\alpha, \beta, \delta)$ can be written as $\psi = A(\alpha)\ B\ (\beta)\ C\ (\delta)$, which is also known as the *product solution*. For illustrating the procedure, we consider Laplace's equation, (1), in rectangular coordinates (although the method is applicable to cylindrical and spherical coordinates also). Thus, from (1) we have

$$\frac{\partial^2\psi}{\partial x^2} + \frac{\partial^2\psi}{\partial y^2} + \frac{\partial^2\psi}{\partial z^2} = 0 \tag{10}$$

Let the solution to (10) be of the form

$$\psi(x,y,z) = X(x)Y(y)Z(z) \tag{11}$$

In (11) notice that the solution is assumed to be a product of three functions: $X(x)$, a function of $x$ alone; $Y(y)$, a function of $y$ only; and $Z(z)$, a function of $z$. Substituting (11) in (10) gives

$$\frac{X''}{X} + \frac{Y''}{Y} + \frac{Z''}{Z} = 0 \tag{12}$$

where $X = X(x)$, $Y = Y(y)$, $Z = Z(z)$, $X'' = \partial^2 X/\partial x^2$, $Y'' = \partial^2 Y/\partial y^2$ and $Z'' = \dfrac{\partial^2 Z}{\partial z^2}$. Notice that (12) must be valid for all values of the variables $x, y$, and $z$. In (12), each term involves one variable. For example, the first term is a function of $x$ only, and the second and third terms do not vary with $x$. Because the first term is the algebraic sum of the second and third terms $[X''/X = -(Y''/Y + Z''/Z)]$, which do not vary with $x$, the first term must also not vary with $x$. This implies that the function $X''/X$ does not very with $x$. This can be true only if the function is a constant, say $k_x^2$. Similarly, the second and third terms are also constants, say $k_y^2$ and $k_z^2$, respectively. We thus have, from (12)

$$k_x^2 + k_y^2 + k_z^2 = 0 \tag{13}$$

The differential equations, involving one variable at a time, are:

$$\frac{d^2X}{dx^2} - k_x^2 X = 0 \tag{14a}$$

$$\frac{d^2Y}{dy^2} - k_y^2 Y = 0 \tag{14b}$$

$$\frac{d^2Z}{dz^2} - k_z^2 Z = 0 \tag{15c}$$

Solutions to these equations are:

$$X = a_x \cosh k_x x + b_x \sinh k_x x \tag{16a}$$

$$Y = a_y \cosh k_y y + b_y \sinh k_y y \tag{16b}$$

$$Z = a_z \cosh k_z Z + b_z \sinh k_z z \tag{16c}$$

where $a's$ and $b's$ are arbitrary constants.

The product of (16a, b and c) is the solution to (10). However, from (13) it is clear that one of the constants must be negative. Let $k_x^2$ be negative (arbitrarily) so that (14a) becomes

$$\frac{d^2X}{dx^2} + k_x^2 X = 0 \tag{17}$$

The solution to (17) becomes

$$X = a'_x \cos k_x x + b'_x \sin k_x x \tag{18}$$

And the complete solution becomes

$$\psi = (a'_x \cos k_x x + b'_x \sin k_x x)\,(a_y \cosh k_y y + \sinh k_y y)$$

$$(\ a_z \cosh k_z z + b_z \sinh k_z z) \tag{19}$$

For two-dimensional cases, it may be shown (see Problem 3.3) that solutions take the forms:

$$\psi = (a_x \cosh k_x x + b_x \sinh k_x x)\,(a'_y \cos k_y y + b'y \sin k_y y) \tag{20}$$

or

$$\psi = (a'_x \cos k_x x + b'_x \sin k_x x)\,(a_y \cosh k_y y + b_y \sinh k_y y) \tag{21}$$

The solution given as (2) is based on the above procedure and the details are given in Problem 3.3.

### 3.5.3 Transform Method of Solving Partial Differential Equations

Linear partial differential equations can also be solved by using Laplace transforms. However, this method is simply an alternative to the methods presented earlier and will only be illustrated by an example.

*Example 3.2*

Solve the wave equation, (3), by Laplace transform subject to the following boundary conditions:

$$u(x,0) = \frac{\partial u}{\partial t}(x,0) = 0$$

$$u(0,t) = f(t)$$

$$\underset{x \to \infty}{Lm}\ u(x,t) = 0$$

*Solution*

Let $U(x,s)$ be the Laplace transform of $u(x,t)$ with respect to the variable $t$, such that

$$L[u(x,t)] \equiv U(x,s) \equiv \int_0^\infty e^{-st} u(x,t)dt \tag{22}$$

Since the given initial conditions are zero,

$$L\left[\frac{\partial^2 u}{\partial t^2}(x,t)\right] = s^2 U(x,s)$$

And, by definition, (22), we also have

$$L\left[\frac{\partial^2 u}{\partial x^2}(x,t)\right] \equiv \int_0^\infty \frac{\partial^2}{\partial x^2}[e^{-st} u(x,t)]dt$$

$$+ \frac{\partial^2}{\partial x^2}\int_0^\infty e^{-st}\ u(x,t)dt = \frac{\partial^2 U}{\partial x^2}(x,s)$$

(subject to the condition that the order of differentiation and integration can be interchanged). Hence, transformed (3) becomes

$$s^2 U(x,s) = c^2 \frac{\partial^2 U}{\partial x^2}(x,s)$$

or,

$$\frac{d^2U}{dx^2} - \frac{s^2}{c^2}U = 0$$

subject to the transformed boundary conditions

$$U(0,s) = F(s) \text{ and } \underset{x \to \infty}{Lm}\ U(x,s) = 0$$

The solution to the transformed equation is

$$U(x,s) = C_1 \exp\left[-\frac{sx}{c}\right] + C_2 \exp\left[\frac{sx}{c}\right]$$

where $C_1$ and $C_2$ may be functions of $s$. Since $s$ is positive, for the second term in this solution to converge, $C_2 = 0$; and $U(0,s) = F(s)$ implies that $C_1 = F(s)$. Thus, the solution to the transformed equation, subject to the given boundary conditions, is

$$U(x,s) = F(s) \exp\left[-\frac{sx}{c}\right]$$

taking inverse transform gives the desired solution

$$u(x,t) = f\left[t - \frac{x}{c}\right]$$

which denotes a forward traveling wave. The term corresponding to the backward wave is implied for $c < 0$.

### 3.6 Classification of Second Order Equations

A general linear second order partial differential equation, in two independent variables, is

$$A\frac{\partial^2 u}{\partial x^2} + B\frac{\partial^2 u}{\partial x \partial y} + C\frac{\partial^2 u}{\partial y^2} + D\frac{\partial u}{\partial x} + E\frac{\partial u}{\partial y} + Fu + G = 0 \tag{23}$$

where $A, B, \ldots, G$ are functions of $x$ and $y$. An alternate form of (23) is

$$A\frac{\partial^2 u}{\partial x^2} + B\frac{\partial^2 u}{\partial x \partial y} + C\frac{\partial^2 u}{\partial y^2} = H\left[x, y, u, \frac{\partial u}{\partial x}, \frac{\partial u}{\partial y}\right] \tag{24}$$

By introducing a suitable change of variables (24) can be expressed in a canonical form. Thus, for the condition $B^2 - 4AC < 0$ the canonical form is

$$\frac{\partial^2 u}{\partial x^2} + \frac{\partial^2 u}{\partial y^2} = H \tag{25}$$

which is called *elliptic equation*. Similarly, for $B^2 - 4AC = 0$, the canonical form is

$$\frac{\partial^2 u}{\partial x^2} = H \tag{26}$$

which is the *parabolic equation*. Finally, when $B^2–4AC > 0$ the canonical form is

$$\frac{\partial^2 u}{\partial x \partial y} = H \tag{27}$$

which is the *hyperbolic equation*.

The above classification aids in choosing a particular numerical method for solving the partial differential equation, as discussed in Chapter 5. Also, the change of variables, leading to the ultimate solution is illustrated by the following example.

*Example 3.3*

Simplify the following hyperbolic equation

$$\frac{\partial^2 u}{\partial x \partial y} + a\frac{\partial u}{\partial x} + b\frac{\partial u}{\partial y} + cu = 0 \tag{1}$$

by the change of variable

$$u(x,y) = w(x,y)e^{-bx - ay} \tag{2}$$

*Solution*

From (2) above

$$\frac{\partial u}{\partial x} = \frac{\partial w}{\partial x} e^{-bx-ay} - bwe^{-bx-ay}$$

$$\frac{\partial u}{\partial y} = \frac{\partial w}{\partial y} e^{-bx-ay} - awe^{-bx-ay}$$

$$\frac{\partial^2 u}{\partial x \partial y} = \frac{\partial^2 w}{\partial x \partial y} e^{-bx-ay} - a\frac{\partial w}{\partial x} e^{-bx-ay} - b\frac{\partial w}{\partial y} e^{-bx-ay} + abwe^{-bx-ay}$$

Substituting these in (1) simplifies it to

$$\frac{\partial^2 w}{\partial x \partial y} + (c - ab)\, w = 0 \tag{3}$$

which can be reduced to Bessel equation by substituting $w(x,y) = X(x)Y(y)$. See Problem 3.5.

In addition to the preceding classification on a mathematical basis engineering problems, involving partial differential equations, are classified as equilibrium, propagation and eigenvalue problems. Respective examples of such problems are: Laplace's equation, wave equation and diffusion equation (See Section 3.4). Such a classification also aids in numerical solutions (Chapter 5).

## Solved Problems

*3.1* Using Newton's law of gravitation, derive Laplace's equation for the gravitational potential $G$ defined by $G = km/r$, where $k$ is a positive constant, and $G$ is the potential at any point $P$ due to a mass $m$ at a distance $r$ from $P$.

*Solution*

$$\text{Since } G = \frac{km}{r} \tag{1}$$

$$\text{grad } G \equiv \nabla G = -\frac{\partial}{\partial r}\left[\frac{km}{r}\right] = -\frac{km}{r^2} = F \tag{2}$$

where $F$ is the force of attraction between a mass $m$ and a unit mass at a distance $r$ from $m$.

$$\text{Now, div (grad } G) = \nabla \cdot \nabla G = \nabla^2 G \tag{3}$$

Thus, from (2) and (3) $\nabla^2 G - 0$, which is Laplace's equation for the gravitational potential.

Alternatively, since $\oint \mathbf{F} \cdot dl = 0$ we must have $\nabla \cdot \mathbf{F} \equiv 0$. Hence the result.

*3.2* Show that the method of separation of variables is applicable to the wave equation, (3), by obtaining two ordinary differential equations.

*Solution*

Let $u(x,t) = X(x)T(t)$

which when substituted in (3) yields

$$\frac{X''}{X} = C^{'h}\frac{T''}{T} = \lambda$$

$$\text{Hence} \quad X'' - \lambda X = 0$$

$$\text{and} \quad T'' - a^2 \lambda T = 0$$

*3.3* For the problem stated in Example 3.1 in the text, show that the solution is correctly given by (2).

*Solution*

Here we have to solve Laplace's equation in two dimensions subject to the given boundary conditions. Proceeding as in Section 3.5.2 and observing that there is periodicity in the $y$-direction, we choose the solution given by (20). The boundary condition $\psi(0,y) = 0$ requires that $a_x = 0$. Also, $\psi(x,0) = 0$ implies that $a'_y = 0$. Furthermore, the relationship $k_x^2 + k_y^2 = 0$ implies that $k_x^2 = -k_y^2$ and the magnitudes of $k_x$ and $k_y$ must be the same; that is, $k_x = k_y = k$, and the form of the solution becomes

$$\psi = C_n \sinh kx \sin ky \tag{1}$$

where $C_n = b_x b'_y$.
Now, from the condition $\psi(x,a) = 0$ we obtain

$$\sin ka = 0 \ \ or \ \ k = \frac{n\pi}{a}, n = \text{an integer}$$

Hence (1) becomes

$$\psi = C_n \sinh \frac{n\pi x}{a} \sin \frac{n\pi y}{a} \tag{2}$$

The boundary condition $\psi(b,y) = \psi_0$ cannot be satisfied by (2). Since $n$ is an integer we will have an infinite number of solutions, and the most general from of the solution becomes

$$\psi = \sum_{n=1}^{\infty} C_n \sinh \frac{n\pi x}{a} \sin \frac{n\pi y}{a} \tag{3}$$

Now, the last mentioned boundary condition yields

$$\psi_0 = \sum_{n=1}^{\infty} a_n \sin \frac{n\pi y}{a} \equiv f(y) \tag{4}$$

$$\text{where } a_n = C_n \sinh \frac{n\pi b}{a} \tag{5}$$

Notice (4) is a Fourier sine series and the Fourier coefficient $a_n$ is given by

$$a_n = 0 \quad n \ \text{ even}$$

$$a_n = \frac{4\psi_0}{n\pi} \quad n \ \text{ odd}$$

Thus, from (5),

$$C_n = \frac{4\psi_0}{n\pi \sinh(n\pi b/a)}, \ n \text{ odd}$$

and the final solution is, therefore, correctly given by (2).
*3.4* Starting with physical laws derive a general equation for heat flow by conduction through a solid of density $\rho$, thermal conductivity $k$ and

specific heat $c_p$.

*Solution*

First, we define the heat flux as the time rate of flow of heat per unit surface area. Then, according to Fourier's law of heat conduction, which states that the heat flux is proportional to the negative temperature gradient, we have

$$\phi = k\frac{\partial u}{\partial n} = k\nabla u \tag{1}$$

where $\phi$ = heat flux, $k$ = thermal conductivity and $u$ = temperature. The heat flux through a closed surface must equal the net gain or loss of heat by the volume enclosed by the surface. Or,

$$\int_s \mathbf{n}\cdot\phi d\mathbf{s} = \int_v \rho c \frac{\partial u}{p\,\partial t}\, dv \tag{2}$$

By divergence theorem, (2) becomes

$$\int_s \mathbf{n}\cdot\phi d\mathbf{s} - \int_v \nabla\cdot\phi dv \tag{3}$$

Combining (1) and (3) gives

$$\int_s \mathbf{n}\cdot\phi d\mathbf{s} = \int_v (\nabla\cdot\mathbf{u}\nabla\mathbf{u})dv = \int_v k\nabla^2 u dv \tag{4}$$

Hence, from (2) and (4) we obtain the general equation for heat flow:

$$\frac{\partial u}{\partial t} = \alpha\nabla^2 u \tag{5}$$

where $\alpha = k/\rho c_p$. Notice that, under steady state, (5) reduces to Laplace's equation. Otherwise, (5) is a form of the diffusion equation.

*3.5* Consider a uniform circular disc of radius $a$ insulated on its lateral surfaces, whose temperature depends only on the distance $r$ from the center of the disc. (a) What is the governing equation which determines the transient temperature distribution in the disk? (b) Using a product solution, obtain two ordinary differential equations from which the complete solution may be obtained.

*Solution*

(a) Equation (5) of Problem 3.3 is the general heat flow equation. In polar coordinates, with no $\theta$-dependence, (5) becomes

$$\frac{\partial^2 u}{\partial r^2} + \frac{1}{r}\frac{\partial u}{\partial r} = \frac{1}{\alpha}\frac{\partial u}{\partial t} \tag{1}$$

which is the required equation.

(b) Let the product solution be $u(r,t) = R(r)T(t)$ which, when substituted in (1), gives

$$\frac{R''}{R} + \frac{1}{r}\frac{R'}{R} = \frac{1}{\alpha}\frac{T'}{T} = \lambda \quad \text{(a constant)} \tag{2}$$

From physical considerations $\lambda$ is negative. Let $\lambda = -\beta^2$. From (2) we obtain

$$T' + \alpha\beta^2\, T = 0$$

$$R'' + \frac{1}{r}R' + \beta^2 R = 0 \tag{4}$$

Equation (4) is known as the *Bessel equation.* If we let $s = \beta r$, (4) becomes

$$\frac{d^2R}{ds^2} + \frac{1}{s}\frac{dR}{ds} + R = 0$$

or,

$$s^2\frac{d^2R}{ds^2} + s\frac{dR}{ds} + s^2R = 0 \tag{5}$$

which is *Bessel equation of zero order.* General *Bessel equation of nth order* is

$$s^2\frac{d^2R}{ds^2} + s\frac{dR}{ds} + (s^2 - n^2)\,R = 0 \tag{6}$$

*3.6* Replace $s$ by $x$ and $R$ by $y$ and obtain a general solution to Bessel equation of nth order:

$$x^2\frac{d^2y}{dx^2} + x\frac{dy}{dx} + (x^2 - n^2)\,y = 0 \tag{1}$$

*Solution*

A power series solution, or the method of Frobenius, can be applied to (1). Thus, let

$$y = x^p \sum_{j=0}^{\infty} a_j x^j = \sum_{j=0}^{\infty} a_j x^{p+j} \tag{2}$$

Substituting (2) in (1) and equating the sum of the coefficients of like powers of $x$ to zero shows that the indicial equation gives $p = \pm\, n$, $a_0$ is arbitrary and $a_1 = 0$. Taking $p = n$, we get

$$a_j = \frac{-a_{j-2}}{j(2n+j)} \qquad j = 2, 3, \ldots$$

of which all coefficients with $j$ = odd are zero. Hence

$$a_{2k} = \frac{(-1)^k a_0}{2^{2k} k!\,(n+1)\,(n+2)\ldots(n+k)} \qquad k = 1,2,\ldots \tag{3}$$

We let $a_0 = 1/[2^n \Gamma(n+1)]$ and recall that $\Gamma$ function $\Gamma(n+k+1) = (n+k)$ $(n+k-1)\ldots(n+1)\Gamma(n+1)$. Thus, we get from (2) and (3)

$$y = J_n(x) \equiv \sum_{k=0}^{\infty} \frac{(-1)_k x^{n+2k}}{2^{n+2k} k! \sqrt{(n+k+1)}} \tag{4}$$

where $J_n(x)$ is the *Bessel function* of the first kind of order $n$. In order to get the complete solution we let $p = -n$ and $a_0 = 2^n/[\Gamma(-n+1)]$ and obtain

$$y = AJ_n(x) + BJ_{-n}(x)$$

If $n$ is an integer then it may be shown that

$$J_{-n}(x) = (-1)^n J_n(x) \tag{5}$$

and

$$J_n(x) = \sum_{k=0}^{\infty} \frac{(-1)^k (x/2)^{n+2k}}{k!\,(n+k)!} \tag{6}$$

*3.7* Electrostatic potential at the surface of a semi-infinite cylinder, of unit radius, is zero, and is $f(r)$ at the base of the cylinder. Also, the potential is zero at infinity. (a) Write the equation governing the potential. (b) Formulate the boundary conditions. (c) Determine the potential subject to the specified boundary conditions.

*Solution*

(a) The governing equation is Laplace's equation in cylindrical coordinates (with no $\theta$-dependence); that is,

$$\frac{\partial^2 \psi}{\partial r^2} + \frac{1}{r}\frac{\partial \psi}{\partial r} + \frac{\partial^2 \psi}{\partial z^2} = 0 \tag{1}$$

(b) Boundary conditions are:

$$\psi(1,z) = \psi(r,\infty) = 0$$

$$\psi(r,0) = f(r)$$

(c) Let the solution be

$$\psi(r,z) = R(r)Z(z)$$

which, when substituted in (1), yields

$$Z'' - \lambda^2 Z = 0 \tag{2}$$

and

$$r^2R'' + rR' + \lambda^2 r^2 R = 0 \tag{3}$$

Solution to (2) is

$$Z = Ae^{-\lambda z} \tag{4}$$

since positive $\lambda$ is not admissible. Equation (3) is Bessel equation of zero order having a solution

$$R = BJ_0(\lambda r) \tag{5}$$

A series solution, for various $\lambda$'s, is possible. Hence,

$$\psi(r,z) = RZ = \sum_{\lambda=0}^{\infty} C_\lambda J_0(\lambda r)\, e^{-\lambda z} \tag{6}$$

Equation (6) will satisfy the condition $\psi(1,z) = 0$ if

$$\psi(r,z) = \sum_{n=1}^{\infty} C_n\, J_0(\alpha_n r)\, e^{-\alpha_n z} \tag{7}$$

where $\alpha_n$ are the roots of $J_0(\alpha) = 0$.
To satisfy $\psi(r,0) = f(r)$ we must have

$$f(r) = \sum_{n=1}^{\infty} C_n J_0(\alpha_n r)$$

where $C_n$ is given by

$$C_n = \frac{2}{[J_1(\alpha_n)]^2} \int_0^1 rf(r)\, J_0(\alpha_n r)\, dr \tag{8}$$

which is the Fourier-Bessel expansion of $f(r)$.

*3.8* Derive the equation governing the motion of a vibrating string. Assume small transverse displacements. The string is subjected to a constant tension $T$ and has a weight $w$ per unit length.

*Solution*

The string and a differential element $ds$ of the string are shown in Figs. 3.2(a) and (b) respectively. Adding the forces in the $y$-direction, from Fig. 3.2(b), we get

$$\sum F_y = -T \sin\alpha + T \sin\beta - wds \tag{1}$$

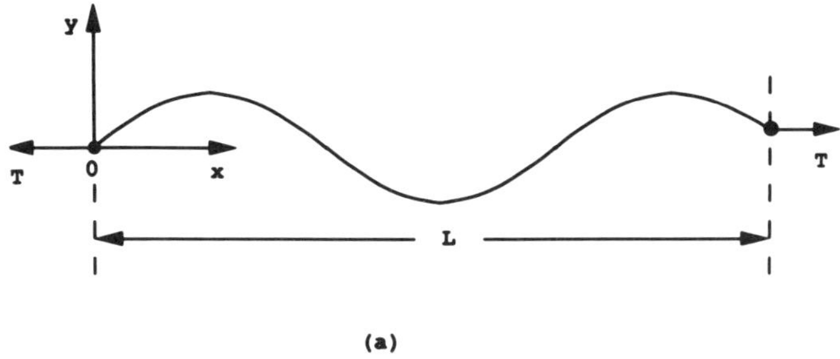


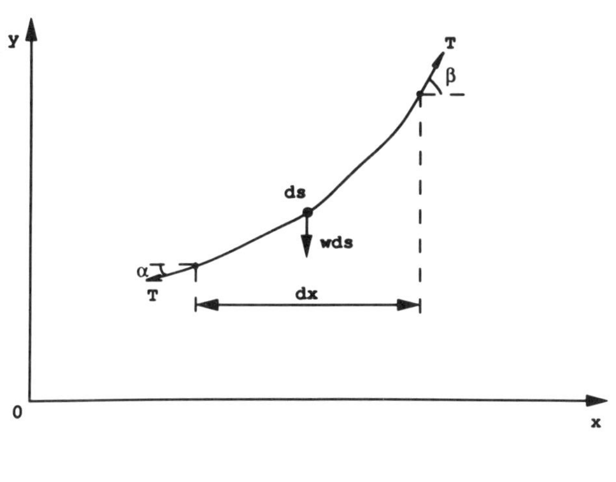


Figure 3.2

Since the displacements are small, the following approximations may be made:

$$\sin\alpha \cong \tan\alpha$$

$$\sin \beta \cong \tan \beta \tag{2}$$

$$ds \cong dx$$

Furthermore, if $u$ is the small displacement

$$\tan \alpha = \frac{\partial u}{\partial x} \tag{3}$$

and

$$\tan \beta \cong \frac{\partial u}{\partial x} + \frac{\partial}{\partial x}\left[\frac{\partial u}{\partial x}\right] dx = \frac{\partial u}{\partial x} + \frac{\partial^2 u}{\partial x^2} dx$$

From (1) - (3), therefore,

$$\sum F_y = T\frac{\partial^2 u}{\partial x^2} - wdx \tag{4}$$

This force must be equal to the gravitational force (which is mass × acceleration). Hence (4) yields

$$T\frac{\partial^2 u}{\partial x^2} - wdx = \frac{w}{g}\, dx\, \frac{\partial^2 u}{\partial t^2}$$

$$a^2 \frac{\partial^2 u}{\partial x^2} = \frac{\partial^2 u}{\partial t^2} + g \tag{5}$$

where $a^2 = gT/w$.

*3.9* Maxwell's equations, in a material medium characterized by $(\mu, \sigma, \varepsilon)$, for electromagnetic fields are stated as:

$$\nabla \times E = -\frac{\partial B}{\partial t} \tag{1}$$

$$\nabla \times H = J + \frac{\partial D}{\partial t} \tag{2}$$

with

$$J = \sigma E \tag{3a}$$

$$B = \mu H \tag{3b}$$

$$D = \varepsilon E \tag{3c}$$

where $E$ = electric field intensity, $D$ = electric flux density, $H$ = magnetic field intensity, $B$ = magnetic flux density and $J$ = electric current density. Obtain a partial differential equation containing B only.

*Solution*
Taking the curl of (2) we get

$$\nabla \times \nabla \times H = \nabla \times \left[ J + \frac{\partial D}{\partial t} \right] \tag{4}$$

Using the vector identity $\nabla \times \nabla \times A = \nabla(\nabla \cdot A) - \nabla^2 A$ and interchanging the order of differentiation in (4) yields

$$\nabla(\nabla \cdot H) - \nabla^2 H = \nabla \times J + \frac{\partial}{\partial t}(\nabla \times D) \tag{5}$$

Since $\nabla \cdot B = 0$, from (3b) $\nabla \cdot H = 0$. Substituting this and (3a) and (3c) in (5) gives

$$-\nabla^2 H = \sigma \nabla \times E + \varepsilon \frac{\partial}{\partial t}(\nabla \times E) \tag{6}$$

Finally, (1), (3b) and (6) can be combined to obtain

$$\nabla^2 H = \mu\sigma \frac{\partial H}{\partial t} + \mu\varepsilon \frac{\partial^2 H}{\partial t^2} \tag{7}$$

*3.10* Modify (7) of Problem (3.9) for (a) free space and (b) for a good conductor.
(a) In free space $\sigma = 0$, $\mu = \mu_0$ and $\varepsilon = \varepsilon_0$. Hence (7) modifies to

$$\nabla^2 H = \mu\varepsilon \frac{\partial^2 H}{\partial t^2}$$

which is the wave equation.
(b) A good conductor is characterized by $\sigma/\varepsilon\omega \gg 1$ such that (7) becomes

$$\nabla^2 H = \mu\sigma \frac{\partial H}{\partial t}$$

which is the diffusion equation.
*3.11* The diffusion equation for the electric field is

$$\nabla^2 E = \mu\sigma \frac{\partial E}{\partial t} \tag{1}$$

At the surface of a semi-infinite conducting slab, shown in Fig. 3.3, the electric field is $E = a_z E_0 \cos \omega t$. Determine the field distribution within the conductor.

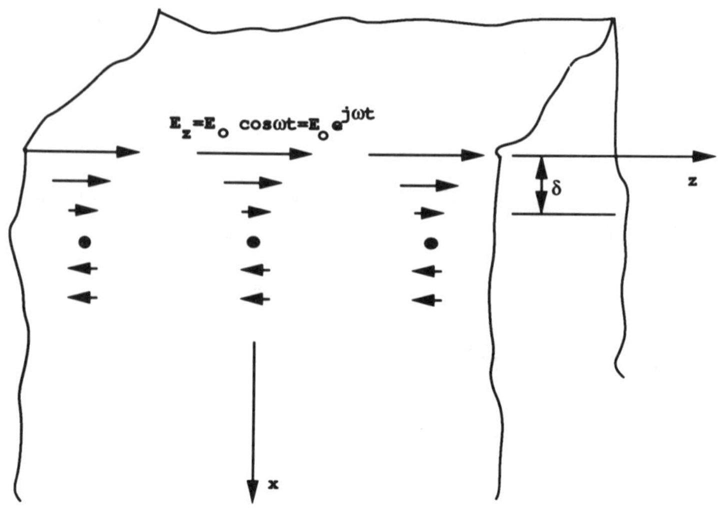


Figure 3.3

*Solution*

Since $E$ is only $z$-directed (1) becomes

$$\frac{\partial^2 E_z}{\partial x^2} = \mu\sigma \frac{\partial E_z}{\partial t} \tag{2}$$

Now, $E_z = E_0 \cos \omega t = E_0 Re(e^{j\omega t}) = E_0 e^{j\omega t}$ (with "Re $\equiv$ real part of" understood). Therefore (2) may be written as

$$\frac{d^2 E_z}{dx^2} = j\omega\mu\sigma E_z = \tau^2 E_z \tag{3}$$

where $\tau^2 \equiv j\omega\mu\sigma$. Observing that $\omega = 2\pi f$ and $\sqrt{j} = (1+j)/\sqrt{2}$ we have

$$\tau = \frac{1}{\delta}(1+j) \tag{4}$$

$$\text{where} \quad \delta = 1/\sqrt{\pi f \mu\sigma} \equiv \textit{skip depth} \tag{5}$$

Solution to (3) becomes

$$E_z = C_1 e^{-\tau x} + C_2^{\tau x} \tag{6}$$

Since $E_z = 0$ at $x = \infty$, $C_2 = 0$. And $E_z = E_0$ at $x = 0$ yields $C_1 = E_0$. Hence, $E_z$ is given by

$$E_z = E_0 e^{-x/\delta} e^{-jx/\delta}$$

which is also sketched in Fig. 3.3

*3.12* A beam has a weight of $w(x,t)$ per unit length. Its Young's modulus is $E$ and $I$ is the moment of inertia of its transverse section. The beam undergoes transverse vibrations $y(x,t)$. Derive the equation governing $y(x,t)$.

*Solution*

If $C$ is the curvature of the beam and $M$ is the bending moment, then from Bernoulli-Euler theory it follows that

$$M = -EIC \tag{1}$$

In terms of derivatives $C$ is given by

$$C = \frac{d^2y/dx^2}{[1 + (dy/dx)]^{3/2}} \tag{2}$$

If the deflection or slope is small, $dy/dx \ll 1$ and (2) becomes

$$C = \frac{\partial^2 y}{\partial x^2} \tag{3}$$

since $y$ is a function of two variables.

Combining (1) and (3) yields

$$M = -EI\,\frac{\partial^2 y}{\partial x^2} \tag{4}$$

where $EI \equiv$ flexural rigidity.

From Newton's second law, the sum of the forces in the $y$-direction, Fig. 3.4(a), is equal to the mass times acceleration:

$$\sum F_y = m\,\frac{\partial^2 y}{\partial t^2} \tag{5}$$

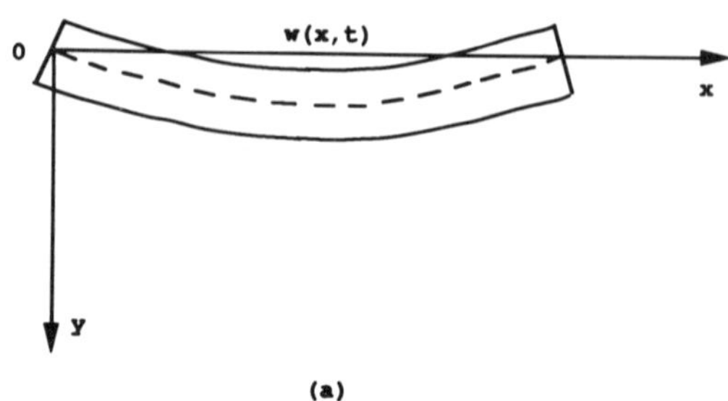


(a)

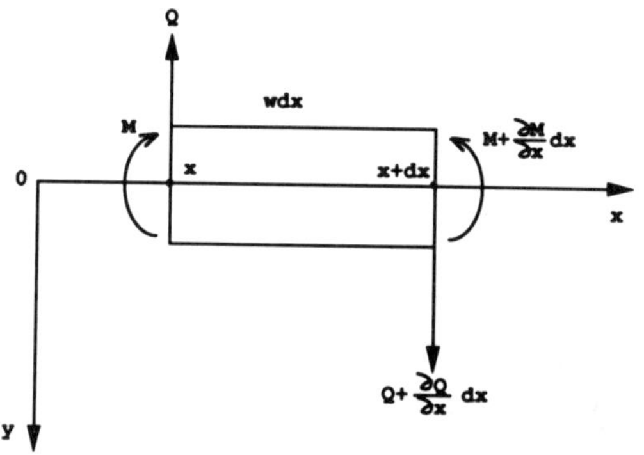


(b)

Figure 3.4

Also, the sum of the external moments about a point is equal to the moment of inertia times angular acceleration (α):

$$\sum M = I\alpha = 0 \tag{6}$$

since the motion is purely transverse and the angular acceleration is zero ($\alpha = 0$).

We now apply (5) and (6) to the element $dx$ of the beam shown in Fig. 3.4(b), where $Q$ is the shear force and $M$ is the moment. For (5), we have

$$-Q + (Q + \frac{\partial Q}{\partial x}\,dx) + wdx = \frac{\partial Q}{\partial x}\,dx + wdx = \frac{wdx}{g}\,\frac{\partial^2 y}{\partial t^2}$$

$$Or, \quad \frac{\partial Q}{\partial x} + w = \frac{w}{g}\,\frac{\partial^2 y}{\partial t^2} \tag{7}$$

Equation (6) requires that

$$M - \left[M + \frac{\partial M}{\partial x}\,dx\right] + (Q + \frac{\partial Q}{\partial x}\,dx)\,dx + (wdx)\,\frac{dx}{2} = 0$$

$$Or, \quad -\frac{\partial M}{\partial x} + Q = 0 \tag{8}$$

where we have neglected the term containing $(dx)^2$.
Eliminating $M$ and $Q$ from (7) and (8) we obtain

$$-\frac{\partial^2}{\partial x^2}\left[EI\,\frac{\partial^2 y}{\partial x^2}\right] + w = \frac{w}{g}\,\frac{\partial^2 y}{\partial t^2} \tag{9}$$

For the special case when $EI$ is constant and the load $w \ll \frac{w}{g}\,\frac{d^2y}{dt^2}$, (9) becomes

$$\frac{\partial^2 y}{\partial t^2} + a^2\,\frac{\partial^4 y}{\partial x^4} = 0 \tag{10}$$

where $a^2 = EI\ g/w$.

*3.13* Solve (10) of Problem 3.12 under the following boundary conditions ($L$ being the length of the beam):

$$y(0,t) = y(L,t) = \frac{\partial^2 y(0,t)}{\partial x^2} = \frac{\partial^2 y}{\partial x^2}\,(L,t) = \frac{\partial y}{\partial t}\,(y,0) = 0$$

and $y(x,0) = Ax(L-x)$.

*Solution*

Let the solution be:

$$y(x,t) = X(x)T(t)$$

This results in the following two ordinary differential equations

$$\frac{d^4X}{dx^4} - \beta^4 X = 0 \tag{1}$$

$$\text{and} \quad \frac{d^2T}{dt^2} + a^2\beta^4 T = 0 \tag{2}$$

where $\beta^4$ is the separation constant.
Solution to (1) may be written as:

$$X(x) = A\ \sinh\beta x + B\ \cosh\beta x + C\ \sin\beta x + D\ \cos\beta x \tag{3}$$

Now, $y(0,t) = X(0)T(t) = 0$ implies that $X(0) = 0$, so that

$$B + D = 0 \tag{4}$$

Next, $y(L,t) = X(L)T(t) = 0$ requires that $X(L) = 0$. Thus,

$$X(L) = A \sinh \beta L + B \cosh \beta L + C \sin \beta L + D \cos \beta L = 0 \tag{5}$$

Also, $\dfrac{d^2X}{dx^2} = A\beta^2 \sinh \beta x + B\beta^2 \cosh \beta x - C\beta^2 \sin \beta x - D\beta^2 \cos \beta x$

Since, $\dfrac{\partial^2 y}{\partial x^2}(0,t) = \dfrac{d^2X}{dx^2}(0)T(t) = 0$ we must have

$$B - D = 0$$

From (4) and (5), $B = D = 0$

The boundy condition $\dfrac{\partial^2 y}{\partial x^2}(L,t) = \dfrac{d^2X}{dx^2}(L)T(t) = 0$ implies that

$$A \sinh \beta L - C \sin \beta L = 0 \text{ (since } B = D = 0) \tag{6}$$

From (5) and (6) we obtain

$$A \sinh \beta L = C \sin \beta L = 0$$

Thus, $A = 0$ (since $\sinh \beta L \neq 0$). We are finally left with the condition

$$C \sin \beta L = 0$$

Since $C \neq 0$, we must have $\beta L = n\pi$ ($n$, an integer) yield the solution to (1):

$$X_n(x) = C_n \sin \frac{n\pi x}{L} \tag{7}$$

The solution to (2) is

$$T(t) = A_1 \cos \alpha\beta^2 t + B_1 \sin \alpha\beta^2 t$$

The boundary condition $\dfrac{\partial y}{\partial t}(x,0) = X(x)\dfrac{dT}{dt}(0) = 0$ requires that $B_1 = 0$. And the solution then becomes

$$T_n(t) = A_n \cos a \frac{n^2\pi^2}{L^2} t \tag{8}$$

The complete solution is, from (7) and (8),

$$y(x,t) = \sum_{n=1}^{\infty} b_n \sin \frac{n\pi}{L} x \cos a \frac{n^2\pi^2}{L^2} t \tag{9}$$

where $b_n = A_n C_n$. For (9) to satisfy $y(x,0) = Ax(L-x)$ we must have

$$Ax(L-x) = \sum_{n=1}^{\infty} b_n \sin \frac{n\pi}{L} x$$

which is a Fourier sine series. Hence

$$b_n = \frac{2}{L}\int_0^L Ax(L-x)\sin\frac{n\pi}{L}x\,dx = \frac{8L^2A}{n^3\pi^3} \qquad n\text{, odd}$$

$$= 0 \qquad n\text{, even}$$

$$\text{Hence } y(x,t) = \frac{8AL^2}{\pi^3}\sum_{n,odd}\frac{1}{n^3}\sin\frac{n\pi}{L}x\cos a\frac{n^2\pi^2}{L^2}t.$$

*3.14* Consider an elementary rectangle, Fig. 3.5, in a thin elastic plate. Normal components of stress $\sigma$ and tangential components of shear $\tau$ are acting on the rectangle, as shown. From the principle of mechanics, derive the biharmonic equation

$$\nabla^4\phi = 0$$

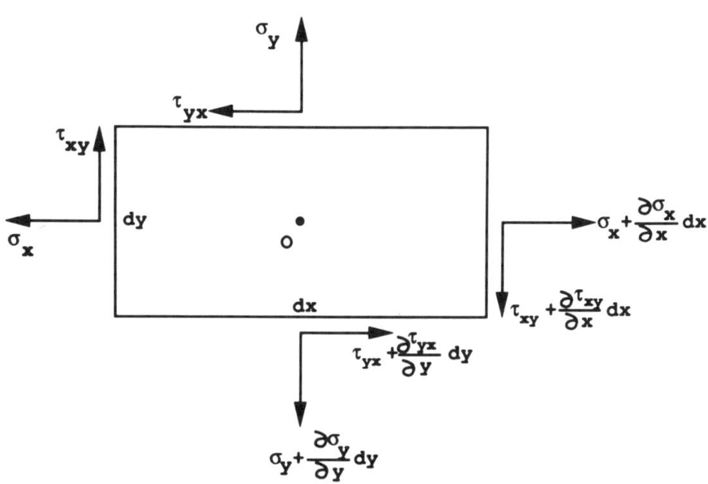


Figure 3.5

where $\phi$ is the Airy stress function such that

$$\sigma_x = \frac{\partial^2\phi}{\partial y^2};\ \sigma_y = \frac{\partial^2\phi}{\partial x^2};\ \text{and } \tau_{xy} = -\frac{\partial^2\phi}{\partial x\,\partial y}$$

*Solution*

Since the element is in equilibrium, we must have $\sum F_x = \sum F_y = 0$. Or

$$\sum F_x = -\sigma_x dy + (\sigma_x + \frac{\partial \sigma_x}{\partial x} dx)\, dy - \tau_{yx}\, dx + \left[\tau_{yx} + \frac{\partial \tau_{yx}}{\partial y} dy\right] dx = 0$$

$$Or, \quad \frac{\partial \sigma_x}{\partial x} + \frac{\partial \tau_{yx}}{\partial y} = 0 \qquad (1)$$

Similarly, from $\sum F_y = 0$ we obtain

$$\frac{\partial \sigma_x}{\partial y} + \frac{\partial \tau_{xy}}{\partial x} = 0 \qquad (2)$$

Now, equating the sum of moments about 0 to zero, we get

$$(\tau_{yx} dx)\frac{dx}{2} + [(\tau_{yx} + \frac{\partial \tau_{yx}}{\partial y} dy)\, dx]\frac{dy}{2}$$

$$-(\tau_{xy} dy)\frac{dx}{2} - [(\tau_{xy} + \frac{\partial \tau_{xy}}{\partial x} dx)\, dy]\frac{dx}{2} = 0$$

Or,

$$\tau_{xy} = \tau_{yx} \qquad (3)$$

where high order infinitesimals have been neglected. Continuity of elastic deformation of the rectangle yields

$$\frac{\partial^2}{\partial x^2}(\sigma_x + \sigma_y) + \frac{\partial^2}{\partial y^2}(\sigma_x + \sigma_y) \qquad (4)$$

which is also known as the equation of compatibility.
We now let

$$\sigma_x = \frac{\partial^2 \phi}{\partial y^2}, \sigma_y = \frac{\partial^2 \phi}{\partial x^2} \text{ and } \tau_{xy} = \tau_{yx} = -\frac{\partial^2 \phi}{\partial x \partial y}$$

which, when used in (1) - (4), yield

$$\frac{\partial}{\partial x}\left[\frac{\partial^2 \phi}{\partial y^2}\right] - \frac{\partial}{\partial y}\left[\frac{\partial^2 \phi}{\partial x \partial y}\right] = 0 \qquad (5)$$

$$\frac{\partial}{\partial y}\left[\frac{\partial^2 \phi}{\partial x^2}\right] - \frac{\partial}{\partial x}\left[\frac{\partial^2 \phi}{\partial x \partial y}\right] = 0 \qquad (6)$$

$$\frac{\partial}{\partial x}\left[\frac{\partial^2 \phi}{\partial y^2}\right] - \frac{\partial}{\partial y}\left[\frac{\partial^2 \phi}{\partial x \partial y}\right] = 0 \qquad (7)$$

$$\frac{\partial}{\partial y}\left[\frac{\partial^2 \phi}{\partial x^2}\right] - \frac{\partial}{\partial x}\left[\frac{\partial^2 \phi}{\partial x \partial y}\right] = 0 \tag{8}$$

Defining $\nabla^4 \equiv \nabla^2(\nabla^2) = \dfrac{\partial^4}{\partial x^4} + 2\dfrac{\partial^4}{\partial x^2 \partial y^2} + \dfrac{\partial^4}{\partial y^4}$, (5) - (8) yield

$$\nabla^4 \phi = 0.$$

## Supplementary Problems

3.15 Starting with the laws of electrostatics, derive Laplace's equation for the electrostatic potential $\Psi$ in a source-free region.

3.16 The equation of motion a horizontal string, fixed at one end and free at the other is,

$$\frac{\partial^2 u}{\partial t^2} = c^2 \frac{\partial^2 u}{\partial x^2} - g$$

where $g$ is acceleration due to gravity. The boundary conditions are:

$$u(x,0) = \frac{\partial u}{\partial t}(x,0) = 0 \quad x > 0$$

$$u(0,t) = 0 \text{ and } \underset{x \to \infty}{Lm} \frac{\partial u}{\partial x}(x,t) = 0$$

Solve the equation by Laplace transform.

*Ans.* $u\ (x,t) = -\dfrac{g}{2c^2}(2cxt - x^2) \quad x \le at$

$$= -\tfrac{1}{2}\, g\, t^2 \quad x \ge at$$

3.17 Verify that $u = A \sin(\omega t + \beta x) + B \sin(\omega t - \beta x)$ is a solution to the wave equation (3). The first term in this solution represents a backward traveling wave and the second term represents a forward traveling wave. What are the velocities of these waves?

*Ans.* $\dot{x} = \pm\ \omega/\beta$

3.18 Solve the wave equation, (3), by separation of variables, subject to the conditions:

$$u(x,0) = f(x);\ \frac{\partial u}{\partial t}(x,0) = \frac{\partial u}{\partial x}(0,t) = \frac{\partial u}{\partial x}(L,t) = 0$$

Ans.

$$u(x,t) = \frac{2}{L} \sum_{n=1}^{\infty} \cos \frac{n\pi}{L} x \cos \frac{n\pi}{L} ct \int_0^L f(\eta) \cos \frac{n\pi}{L} \eta \, d\eta$$

3.19 Diffusion of heat in an infinitely long cylinder, of radius $a$, is given by

$$\alpha \left[ \frac{\partial^2 u}{\partial r^2} + \frac{1}{r} \frac{\partial u}{\partial r} \right] = \frac{\partial u}{\partial t}$$

The surface of the cylinder is maintained at zero temperature and the initial temperature is a function of $f$; that is $u(r,0) = f(r)$. What is the transient temperature distribution within the cylinder?
Ans.

$$u(r,t) = \sum_{n=1}^{\infty} C_n J_0(\lambda_n r) e^{-\lambda_n^2 \alpha t}$$

where $C_n$ is given by the Fourier-Bessel expansion. [See (8) of Problem 3.7].
3.20 In Problem 3.8, let $T \gg w$ such that $g \ll \partial^2 u / \partial t^2$. Subject to this assumption and the following boundary conditions, solve (5) of Problem 3.8. Notice that (5) is the wave equation. The boundary conditions are:

$$u(0,t) = u(L,t) = 0; \ \frac{\partial u}{\partial t}(x,0) = \text{ and } u(x,0) = f(x)$$

where $L$ = length of the string.
Ans.

$$u(x,t) = \frac{2}{L} \sum_{n=1}^{\infty} \left[ \int_0^L f(\eta) \sin \frac{n\pi}{L} \eta \, d\eta \right] \sin \frac{n\pi}{L} x \cos \frac{n\pi a}{L} t$$

3.21 Proceeding as in Problem 3.8 show that the displacement $u(x,y,t)$ of a vibrating thin membrane is given by

$$a^2 \left[ \frac{\partial^2 u}{\partial x^2} + \frac{\partial^2 u}{\partial y^2} \right] = \frac{\partial^2 u}{\partial t^2}$$

where $a^2 = g\, s/w$, g = acceleration due to gravity, $w$ = weight of the membrane/unit area, and $s$ = force per unit length of the membrane, normal to its boundary. Assume that the total force on boundary is much greater compared to the weight of the membrane.

3.22 A bar of square cross-section is made of homogeneous, isotropic and elastic material (satisfying Hooke's law: $\sigma = \varepsilon E$, where $\sigma$ = stress, $E$ = Young's modulus, and $\varepsilon$ = unit strain). The bar undergoes longitudinal vibrations. Assuming small displacements compared to the dimensions of the bar, show that its motion is given by

$$a^2 \frac{\partial^2 u}{\partial x^2} = \frac{\partial^2 u}{\partial t^2}$$

where $a^2 = gE/w$, $w$ being the weight per unit volume of the bar.

*Note*: $\varepsilon \equiv$ unit strain $\equiv \dfrac{(\text{new length} - \text{old length})}{\text{old length}}$

3.23 The cross-section of a semi-infinite plate of permeability $\mu$ is shown in Fig. 3.6. Within the plate the magnetic field $H$ is only $z$-directed and varies with $y$ only.

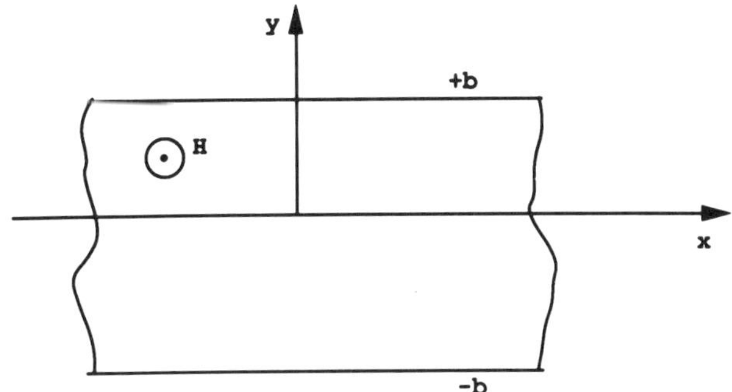


Figure 3.6

(a) Derive the equation governing $H$ in the plate.

(b) For a sinusoidal time variation, modify this equation.

(c) Define

$\phi_s \equiv \int_{-b}^{b} \mu H \, dy \equiv$ flux per unit width of the plate, and determine the H-field distribution within the plate.

Ans.

(a) Diffusion equation

(c) $H = A \cosh \alpha y / \sinh \alpha b$

where $A = \alpha \phi_s / 2\mu$ and $\alpha = (1 + j)/\delta$

[ $\delta$ = skin depth = $1/\sqrt{\pi f \mu \sigma}$]

3.24 In Problem 3.12, in obtaining (9) from (7) and (8) certain steps have been skipped. Complete these steps and verify that (9) follows from (7) and (8).

CHAPTER

4

# DIFFERENCE EQUATIONS

## 4.1 The Linear, Constant-Coefficient Difference Equation

An ordinary, differential equation considered in Chapter 2 relates a continuous function of one variable, *y(t)*, and its various derivatives to a known forcing function, *f(t)*. A difference equation, on the other hand, relates a sequence of numbers or a discrete function of one or more variables to another (known) sequence or discrete function. For example, the ordinary differential equation

$$\frac{dy(t)}{dt} + a_0 y(t) = f(t) \tag{1}$$

can be discretized by noting that

$$\frac{dy(t)}{dt} = \frac{y(t + \Delta t) - y(t)}{\Delta t} \tag{2}$$

where $\Delta t$ is some suitably small increment of $t$. Equation (1) then becomes, approximately,

$$y(t+\Delta t) = (1 - a_0\Delta t)\, y(t) + \Delta t\, f(t) \tag{3}$$

If we choose to solve (3) at discrete increments of $t = k\Delta t$ with $k = 0,1$ ..., then we obtain

$$\begin{aligned}
y(\Delta t) &= (1 - a_0\Delta t)\, y(0) + \Delta t\, f(0) \\
y(2\Delta t) &= (1 - a_0\Delta t)\, y(\Delta t) + \Delta t\, f(\Delta t) \\
y(3\Delta t) &= (1 - a_0\Delta t)\, y(2\Delta t) + \Delta t\, f(2\Delta t) \\
&\vdots \\
y((k+1)\Delta t) &= (1 - a_0\Delta t)\, y(k\Delta t) + \Delta t\, f(k\Delta t)
\end{aligned} \tag{4}$$

Given an initial condition, $y(1)$, we can solve the first equation in (4) for $y(\Delta t)$. Substituting this result into the second equation in (4) we may solve for $y(2\Delta t)$. This procedure can be continuously used to solve, "recursively", for $y$ at $\Delta t$ increments of $t$.

There are many other instances in which similar recursion relations or difference equations occur. The majority are linear with constant coefficients which is the type considered in this chapter. The general form of the *n-th order, linear, constant coefficient, difference equation* is

$$y(k+n) + a_{n-1}\, y(k+n-1) + \ldots + a_1 y(k-1) + a_0 y(k) = f(k) \tag{5}$$

Essentially, a difference equation is simply a relation between members of one sequence of numbers, $y(k)$, and members of another sequence of numbers, $f(k)$. Here the sequence ..., y(-2), y(-1), y(0), y(2), ... is related to the "forcing sequence" ..., f(-2), f(-1), f(0), f(1), f(2), ... by (5). Other members of the forcing sequence may be added to the right-hand side of (5) but this adds no generality. For example, $f(k) + 3f(k+2)$ may be easily combined into one number since *f(k)* and *f(k+2)* are presumed known.

The equation in (5) is said to be constant coefficient if none of the $a_i$ coefficients are functions of *k*. In other words the relation does not depend on where we are in the sequence. The equation $y(k+1) + 2k\, y(k) = f(k)$ depends on $k$, e.g. for *k=1*, $y(2) + 2y(1) = f(1)$ whereas for $k=4$, $y(5) + 8y(4) = f(4)$ and the relationship changes with $k$. A linear difference equation is one in which *y(i)* is not raised to a power greater than unity and no products of *y(i)* and *y(i+j)* appear. As in the case of a linear, ordinary differential equation, a linear difference equation obeys the property of superposition; that is, if $y = y_1$ satisfies (5) with $f = f_1$, $y = y_2$ satisfies (5) with $f = f_2$, ..., $y = y_m$ satisfies (5) with $f = f_m$, then $y = y_1 + y_2 + ... + y_m$ satisfies (5) when $f = f_1 + f_2 + ... + f_m$. The reader should observe the similarities of these equations and their properties to those of the $n$-th order, linear, constant coefficients, ordinary differential equation of Chapter 2. In fact, there is a one-to-one correspondence between the techniques of solution for those equations and the difference equations of this chapter.

## 4.2 Homogeneous and Particular Solutions

Again we may add the "zero sequence forcing function" to the right-hand side of (5):

$$y(k+n) + a_{n-1}\, y(k+n-1) + ... + a_1 y(k-1) + a_0 y(k) = 0 + f(k) \quad (6)$$

Consequently, the solution to (6) will, by superposition, consist of the homogeneous solution, $y_h(k)$;

$$y_h(k+n) + a_{n-1}\, y_h(k+n-1) + ... + a_1 y_h(k-1) + a_0 y_h(k) = 0 \quad (7)$$

and the particular solution, $y_p(k)$:

$$y_p(k+n) + a_{n-1}\, y_p(k+n-1) + ... + a_1 y_p(k-1) + a_0 y_p(k) = f(k) \quad (8)$$

as

$$y(k) = y_h(k) + y_p(k) \quad (9)$$

## 4.3 The Homogeneous Solution

The solution to the homogeneous equation in (7) is of the form

$$y_h(k) = r^k \quad (10)$$

Substituting in (7) yields

$$(r^n + a_{n-1} r^{n-1} + ... + a_1 r + a_0)\, r^k = 0 \quad (11)$$

from which we observe that there are $n$ roots $r_1, r_2 ..., r_n$ of the characteristic equation

$$r^n + a_{n-1} r^{n-1} + ... + a_1 r + a_0 = 0 \quad (12)$$

Therefore the homogeneous solution will consist of the linear combination of "eigenfunctions", $\phi_i(k)$, as

$$y_h(k) = C_1\phi_1(k) + C_2\phi_2(k) + \ldots + C_n\phi_n(k) \tag{13}$$

where the eigenfunctions are

$$\phi_1(k) = r_1^k,\ \phi_2(k) = r_2^k,\ \ldots,\ \phi_n(k) = r_n^k \tag{14}$$

(a) *Real, repeated roots*

Suppose one of the roots, $r_i$, of the characteristic equation is of multiplicity $m$, i.e., $r_i = r_{i+1} = \ldots r_{i+m-1}$, then the eigenfunctions associated with this root are

$$\phi_i = r_i^k,\ \phi_{i+1} = k\phi_i = kr_i^k,\ \ldots,\ \phi_{i+m-1} = k^{m-1}\phi_i = k^{m-1}r_i^k \tag{15}$$

(b) *Complex roots*

Suppose one of the roots of the characteristic equation is complex with its associated conjugate root:

$$r_i = \alpha_i + j\beta_i,\ r_{i+1} = \alpha_i - j\beta_i \tag{16}$$

The portion of the homogeneous solution corresponding to these roots becomes

$$C_1(\alpha_i + j\beta_i)^k + C_{i+1}(\alpha_i - j\beta_i)^k \tag{17}$$

Noting that

$$(\alpha_i + j\beta_i)^k = \rho_i^k e^{j\theta_i k} \tag{18}$$

where

$$\rho_i = \sqrt{\alpha_i^2 + \beta_i^2} \tag{19}$$

and

$$\theta_i = \tan^{-1}\frac{\beta_i}{\alpha_i} \tag{20}$$

(17) can be written as

$$\begin{aligned} &C_i\rho_i^k e^{j\theta_i k} + C_{i+1}\rho_i^k e^{-j\theta_i k} \\ &= \rho_i^k[(C_i + C_{i+1})\cos(\theta_i k) + j(C_i - C_{i+1})\sin(\theta_i k)] \end{aligned} \tag{21}$$

and the eigenfunctions may be chosen as

$$\phi_i = \rho_i^k\cos(\theta_i k),\ \phi_{i+1} = \rho_i^k\sin(\theta_i k) \tag{22}$$

(c) *Repeated, complex roots*
Suppose the complex root in (16) is of multiplicity $m$. Then the corresponding eigenfunctions are

$$\begin{aligned} \phi_i &= \rho_i^k \cos(\theta_i k),\ \phi_{i+1} = \rho_i^k \sin(\theta_i k) \\ \phi_{i+2} &= k\phi_i,\quad \phi_{i+3} = k\phi_{i+1} \\ &\vdots \qquad\qquad \vdots \\ \phi_{i+2m-2} &= k^{m-1}\theta_i,\quad \phi_{i+2m-1} = k^{m-1}\theta_{i+1} \end{aligned} \tag{23}$$

*Example 4-1*
Determine a solution of the difference equation

$$y(k+2) - 4y(k+1) + 4y(k) = 0$$

*Solution*
The characteristic equation is

$$r^2 - 4r + r = 0$$

which has roots $r_1 = -2$, $r_2 = -2$. The homogeneous solution is

$$y_h(k) = C_1(-2)^k + C_2 k(-2)^k$$

*Example 4-2*
Suppose the roots of a 6-th order characteristic equation are $r_1 = 4$, $r_2 = 3$, $r_3 = 1+j2$, $r_4 = 1-j2$, $r_5 = 1+j2$, $r_6 = 1-j2$. Write the form of the homogeneous solution.
*Solution*

$$\begin{aligned} y_h(k) &= C_1 4^k + C_2 3^k + C_3(\sqrt{5})^k \cos(k\theta) + C_4(\sqrt{5})^k \sin(k\theta) \\ &+ C_5 k(\sqrt{5})^k \cos(k\theta) + C_6 k(\sqrt{5})^k \sin(k\theta) \end{aligned}$$

where

$$\theta = \tan^{-1} 2$$

## 4.4 Linear Independence of Discrete Functions

In a manner similar to continuous functions, the set of discrete functions $\phi_1(k)$, $\phi_2(k)$, ..., $\phi_n(k)$ is said to be linearly independent for $k_1 \le k \le k_2$ if

$$C_1\phi_1(k) + C_2\phi_2(k) + \ldots + C_n\phi_n(k) = 0 \tag{24}$$

and only if $C_1 = C_2 = ... = C_n = 0$. Basically this means that none of the functions can be obtained as a linear combination of the remaining functions of the set; that is, if it is not possible to find a set of constants such that

$$\phi_i(k) = C_1\phi_1(k) + ... + C_{k-1}\phi_{i-1}(k) + C_{i+1}\phi_{i+1}(k) + ... + C_n\phi_n(k) \quad (25)$$

A convenient technique for determining the linear independence of a set of discrete functions is the following. Assuming (24) to be true we may generate the following set of $n$ equations in $C_1, C_2, ..., C_n$:

$$\begin{aligned} C_1\phi_1(0) + C_2\phi_2(0) + ... + C_n\phi_n(0) &= 0 \\ C_1\phi_1(1) + C_2\phi_2(1) + ... + C_n\phi_n(1) &= 0 \\ &\vdots \\ C_1\phi_1(n-1) + C_2\phi_2(n-1) + ... + C_n\phi_n(n-1) &= 0 \end{aligned} \quad (26)$$

The functions will be linearly independent if and only if the solution to (26) is $C_1 = C_2 ... = C_n = 0$ which translates, using Cramer's rule, to the requirement that the following determinant be nonzero:

$$C = \begin{vmatrix} \phi_1(0) & \phi_2(0) & ... & \phi_n(0) \\ \phi_1(1) & \phi_2(1) & ... & \phi_n(1) \\ \vdots & \vdots & & \vdots \\ \phi_1(n) & \phi_2(n) & ... & \phi_n(n) \end{vmatrix} \neq 0 \quad (27)$$

The determinant in (27) is called the Casorati and its nonzero value insures the linear independence of the $\phi_i$. The eigenfunctions of the homogeneous solution, $\phi_i(k)$, when chosen according to the rules of the previous section, will be linearly independent.

## 4.5 The Particular Solution

As in the case of ordinary differential equations, a simple technique for obtaining the particular solution of (8) is the method of undetermined coefficients. Also if the right-hand side of (8) is the sum of $m$ forcing functions:

$$y_p(k+n) + a_{n-1}\, y_p(k+n-1) + ... + a_0\, y_p(k) = f_1(k) + f_2(k) + ... + f_m(k) \quad (28)$$

then the particular solution will consist, by superposition, of the sum of the particular solutions to (28) with the right-hand side containing each $f_i(k)$ individually:

$$y_{p1}(k+n) + a_{n-1}\, y_{p1}(k+n-1) + \ldots + a_0 y_{p1}(k) = f_1(k)$$
$$y_{p2}(k+n) + a_{n-1}\, y_{p2}(k+n-1) + \ldots + a_0\, y_{p2}(k) = f_2(k) \qquad (29)$$
$$\vdots$$
$$y_{pm}(k+n) + a_{n-1}\, y_{pm}(k+n-1) + \ldots + a_0\, y_{pm}(k) = f_m(k)$$

as

$$y_p(k) = y_{p1}(k) + y_{p2}(k) + \ldots + y_{pm}(k) \qquad (30)$$

Table 4–1

| $f_i(k)$ | Form of $y_p(k)$ |
|---|---|
| $M\ \rho^k$ | $K\ \rho^k$ |
| $M\ \sin\theta k$ or $M\ \cos\theta k$ | $K_{c0}\cos\theta k + K_{s0}\sin\theta k$ |
| $M\ \rho^k\ k^{\ddagger}$ | $\rho^k\ \{K_0 + K_1 k + \ldots + K_{\ddagger} k^{\ddagger}\}$ |
| $M\ \rho^k\ \sin\theta k$ or $M\ \rho^k\ \cos\theta k$ | $\rho^k\ \{K_{c0}\cos\theta k + K_{s0}\sin 0k\}$ |

The reader should compare these forms to the analogous forms for ordinary differential equations given in Table 2-1 of Chapter 2. The analogous quantities are

| $f(k)$ | | $f(t)$ |
|---|---|---|
| $\rho^k$ | $\rightarrow$ | $e^{\alpha t}$ |
| $\sin\theta k$ | $\rightarrow$ | $\sin\beta t$ |
| $\cos\theta k$ | $\rightarrow$ | $\cos\beta t$ |
| $k^{\ddagger}\rho^k$ | $\rightarrow$ | $t^{\ddagger} e^{\alpha t}$ |

A general result can be obtained by writing

$$f_k(k) = \begin{cases} M\ k^{\ddagger}\rho^k\ \cos\theta k \\ \quad or \\ M\ k^{\ddagger}\rho^k\ \sin\theta k \end{cases} \qquad (31)$$

For both these forms we would assume

$$y_p(k) = \rho^k\ \cos\theta k\ \{k_{c0} + k_{c1}k + \ldots + k_{c\ddagger} k^{\ddagger}\} \qquad (32)$$
$$+ \rho^k\ \sin\theta k\ \{k_{s0} + k_{s1}k + \ldots + k_{s\ddagger} k^{\ddagger}\}$$

As is the case with ordinary differential equations, this method will fail if any of the components of the assumed forms, $y_p(k)$, is of the same form as any of the eigenfunctions, $\phi_j(k)$, in the homogeneous solution: a result which is clear since $\phi_k(k)$ satisfy the homogeneous equation and cannot satisfy the inhomogeneous equation. To remedy this problem, we simply multiply the usual assumed form of $y_p(k)$ in (32) by a power of $k$, $k^m$, such that the assumed form no longer contains a factor of the form of any of the eigenfunctions $\phi_j(k)$.

*Example 4-3*

Consider the difference equation

$$y(k+2) - 4y(k+1) + 4y(k) = 3k^2 + 4^k + 5 \cdot 2^k$$

Determine a complete solution.

*Solution*

The characteristic equation is

$$r^2 - 4r + 4 = 0$$

which as roots $r_1 = 2, r_2 = 2$. Thus the homogeneous solution is

$$y_h(k) = C_1 2^k + C_2 k 2^k$$

The particular solution can be obtained by writing the difference equation as

$$y(k+2) - 4y(k+1) + 4y(k) = f_1(k) + f_2(k) + f_3(k)$$

where we may identify

$$f_1(k) = 3k^2$$

$$f_2(k) = 4^k$$

$$f_2(k) = 5 \cdot 2^k$$

Corresponding to $f_1(k) = 3k^2$ we assume

$$y_{p1}(k) = K_0 + K_1 k + K_2 k^2$$

Substituting we obtain

$$y_{p1}(k+2) - 4y_{p1}(k+1) + 4y_{p1}(k) = K_0 + K_1(k+2) + K_2(k+2)^2$$

$$-4K_0 - 4K_1(k+1) - 4K_2(k+2)^2$$

$$+4K_0 + 4K_1 k + 4K_2 k^2$$

$$= (K_0 - 2K_1 - 12K_2) + (K_1 - 12K_2)k$$
$$+ K_2k^2$$
$$= 3k^3$$

which becomes, by matching powers of $k$,

$$K_2 = 3$$
$$K_1 - 12K_2 = 0$$
$$K_0 - 2K_1 - 12K_2 = 0$$

which may be solved as $K_2 = 3$, $K_1 = 36$, $K_0 = 108$ so that

$$y_{p1}(k) = 108 + 36k + 3k^2$$

Corresponding to $f_2(k) = 4^k$ we assume

$$y_{p2}(k) = K\,4^k$$

and substituting yields

$$K\,4^{(k+2)} - 4K\,4^{(k+1)} + 4K\,4^k = 4K\,4^k$$
$$= 4^k$$

so that $K = 1/4$. Thus

$$y_{p2}(k) = 1/4 \cdot 4^k$$

For $f_3(k) = 5{\cdot}2^k$, we would normally assume $y_{p3}(k) = K\,2^k$ but the homogeneous solution contains a term of this form. Multiplying by $k$ yields $y_{p3}(k) = K\;k\,2^k$ but the homogeneous solution also contains a term of this form so we once again multiply by $k$ to obtain

$$y_{p3}(k) = K\;k^2\,2^k$$

This form is not contained in the homogeneous solution. Substitution yields

$$K\,(k+2)^2\,2^{(k+2)} - 4\,K\,(k+1)^2\,2^{(k+1)} + 4\,K\;k^2\,2^k = 8\;K\;2^k$$
$$= 5{\cdot}2^k$$

so that $K = 5/8$. Thus the form of the solution is

$$y(k) = y_h(k) + y_p(k)$$
$$= C_1 2^k + C_2 k\;2^k + 108 + 36k + 3k^2 + 1/4 \cdot 4^k + \frac{5}{8}\,k^2\,2^k$$

*Example 4-4*

Suppose the difference equation in the previous example has $5k \cdot 2^k$ as the forcing function:

$$y(k+2) - 4y(k+1) + 4y(k) = 5k\ 2^k$$

Write the form of the solution.

*Solution*

The homogeneous solution is

$$y_h(k) = C_1 2^k + C_2 k\ 2^k$$

since $r_1 = r_2 = 2$ is a repeated root of multiplicity 2. The assumed form of the particular solution would normally be

$$y_p(k) = 2^k\ \{K_0 + K_1 k\}$$
$$= K_0 2^k + K_1 k 2^k$$

However the homogeneous solution contains these factors so we must multiply the usual form by $k^2$:

$$y_p(k) = K_0 k^2\ 2^k + K_1 k^3\ 2^k$$

and this result does not contain a term contained in the homogeneous solution. Substituting the assumed form yields

$$K_0(k+2)^2\ 2^{(k+2)} + K_1(k+2)^3\ 2^{(k+2)} - 4\ K_0(k+1)^2\ 2^{(k+1)} - 4K_1(k+1)^3\ 2^{(k+1)}$$
$$+ 4\ K_0 k^2\ 2^k + 4\ K_1\ k^3\ 2^k = 5k\ 2^k$$

Collecting powers of $k$ we obtain $K_0 = -5/8$, $K_1 = 5/24$ so that

$$y_p(k) = -\frac{5}{8}\ k^2\ 2^k + \frac{5}{24}\ k^3\ 2^k$$

Thus

$$y(k) = y_h(k) + y_p(k)$$
$$= C_1 2^k + C_2 k 2^k - \frac{5}{8}\ k^2\ 2^k + \frac{5}{24}\ k^3 2^k$$

## 4.6 The Total Solution

The complete solution is, by superposition, the sum of the homogeneous and particular solutions:

$$y(k) = C_1 \phi_1(k) + \ldots + C_n \phi_n(k) + y_p(k) \qquad (33)$$

In order to evaluate the undetermined constants $C_1, ..., C_n$ in (33) we must prescribe additional conditions. The difference equation in (5) relates $y(k+n)$ (and all higher-order differences) to $y(k+n-1), ..., y(k)$. Thus we may independently prescribe $n-1$ conditions. For example, we may prescribe $y(0), y(1), ..., y(n-1)$. Applying these to (33) we obtain

$$\begin{aligned} C_1\phi_1(0) + ... + C_n\phi_n(0) &= y(0) - y_p(0) \\ C_1\phi_1(1) + ... + C_n\phi_n(1) &= y(1) - y_p(1) \\ &\vdots \\ C_1\phi_1(n-1) + ... + C_n\phi_n(n-1) &= y(n-1) - y_p(n-1) \end{aligned} \tag{34}$$

and in matrix notation, (34) becomes

$$\begin{bmatrix} \phi_1(0) & \phi_2(0) & ... & \phi_n(0) \\ \phi_1(1) & \phi_2(1) & ... & \phi_n(1) \\ \vdots & \vdots & & \vdots \\ \phi_1(n-1) & \phi_2(n-1) & ... & \phi_n(n-1) \end{bmatrix} \begin{bmatrix} C_1 \\ C_2 \\ \vdots \\ C_n \end{bmatrix} = \begin{bmatrix} y(0) & - & y_p(0) \\ y(1) & - & y_p(1) \\ & \vdots & \\ y(n-1) & - & y_p(n-1) \end{bmatrix} \tag{35}$$

which may be solved for $C_1, C_2, ..., C_n$ by, for example, Cramer's rule or any of the other techniques of Chapter 1. The determinant of the coefficient matrix, $|\mathbf{C}|$, is the Casorati discussed in Section 4-4. Since the $\phi_i(k)$ eigenfunctions of the homogeneous solution were chosen such that they are linearly independent, the coefficient matrix is non-singular thus guaranteeing the unique solution for $C_1, C_2, ..., C_n$.

*Example 4-5*

Determine the solution to

$$y(k+2) - 5y(k+1) + 6y(k) = 5^k$$

subject to the conditions $y(0) = 1, y(1) = 0$.

*Solution*

The characteristic equation is

$$r^2 - 5r + 6 = 0$$

with roots $r_1 = 2, r_2 = 3$. Thus

$$y_h(k) = C_1 2^k + C_2 3^k$$

The assumed form of the particular solution is $y_p(k) = K \cdot 5^k$. Substitution yields $K = 1/6$ so that

$$y(k) = C_1\, 2^k + C_2\, 3^k + \frac{1}{6} \cdot 5^k$$

Applying the initial conditions yields

$$C_1 + C_2 + \frac{1}{6} = 1$$

$$C_1 \cdot 2 + C_2 \cdot 3 + \frac{5}{6} = 0$$

which may be written as

$$\begin{bmatrix} 1 & 1 \\ 2 & 3 \end{bmatrix} \begin{bmatrix} C_1 \\ C_2 \end{bmatrix} = \begin{bmatrix} 5/6 \\ -5/6 \end{bmatrix}$$

Solving we obtain $C_1 = 10/3$, $C_2 = -5/2$ and the total solution is

$$y(k) = \frac{10}{3} \cdot 2^k - \frac{5}{2} \cdot 3^k + \frac{1}{6} \cdot 5^k$$

## 4.7 Particular Solution for Sinusoidal Forcing Function

If $f_i(k)$ is of the form

$$f_i(k) = \begin{cases} M\rho^k \cos \theta k \\ or \\ M\rho^k \sin \theta k \end{cases} \tag{36}$$

We normally assume a form of the particular solution to be

$$y_p(k) = K_{c0}\, \rho^k \cos \theta k + K_{s0}\, \rho^k \sin \theta k \tag{37}$$

A simpler technique which is analogous to the method of Section 2-7 for ordinary differential equations is the following. Instead of $f_i(k)$ in (36) we obtain the particular solution, $\hat{y}_p(k)$ with

$$f_i(k) = M\rho^k e^{j\theta k} \tag{38}$$

If order to solve for $y_p(k)$ we assume a form

$$\hat{y}_p(k) = K\, \rho^k\, e^{j\theta k} \tag{39}$$

Substitution yields

$$k\rho^{(k+n)}\, e^{j\theta(k+n)} + a_{n-1}\, K\rho^{(k+n-1)}\, e^{j\theta(k+n-1)} + \ldots + a_0 K\rho^k\, e^{j\theta k} \tag{40}$$
$$= M\rho^k d^{j\theta k}$$

or

$$K = \frac{M}{\rho^n e^{j\theta n} + a_{n-1}\rho^{(n-1)} e^{j\theta(n-1)} + \ldots + a_0} \tag{41}$$

$$= |K| \angle \theta_K$$

The desired solution is found from

$$\begin{array}{ll} \underline{f_i(k)} & \underline{y_p(k)} \\ M\rho^k \cos\theta k & \mathrm{Re}\,\{\hat{y}_p(k)\} \\ M\rho^k \sin\theta k & \mathrm{Im}\,\{\hat{y}_p(k)\} \end{array} \tag{42}$$

The proof of this result is rather simple. Since $e^{j\theta k} = \cos\theta k + j\sin\theta k$, the solution $\hat{y}_p(k)$ is the sum of the solution due to $\cos\theta k$ and $j$ times the solution due to $\sin\theta k$. By superposition, (42) results.

*Example 4-5*

Determine the particular solution to

$$y(k+2) + y(k+1) + y(k) = 5 \cdot 2^k \cos 6k$$

*Solution*

By the usual method we assume

$$y_p(k) = 2^k \{K_{c0}\cos 6k + K_{s0}\sin 6k\}$$

Substitution yields

$$2^{(k+2)}\{K_{c0}\cos 6(k+2) + K_{x0}\sin 6(k+2)\}$$

$$+ 2^{(k+1)}\{K_{c0}\cos 6(k+1) + K_{s0}\sin 6(k+1)\}$$

$$+ 2^k\{K_{c0}\cos 6k + K_{s0}\sin 6k\} = 5 \cdot 2^k \cos 6k$$

Using $\cos(A+B) = \cos A\cos B - \sin A\sin B$ and $\sin(A+B) = \sin A\cos B + \cos A\sin B$ and grouping coefficients of $\cos 6k$ and $\sin 6k$ we obtain

$$\begin{bmatrix} (4\cos 12 + 2\cos 6 + 1) & (4\sin 12 + 2\sin 6) \\ (-4\sin 12 - 2\sin 6) & (4\cos 12 + 2\cos 6 + 1) \end{bmatrix} \begin{bmatrix} K_{c0} \\ K_{s0} \end{bmatrix} = \begin{bmatrix} 5 \\ 0 \end{bmatrix}$$

which may be solved for $K_{c0} = .67$ and $K_{s0} = -.29$. Thus

$$y_p(k) = 2^k\{.67\cos 6k - .29\sin 6k\}$$

Using the alternate method of this section we obtain

$$K = \frac{5}{2^2 e^{j12} + 2e^{j6} + 1}$$

$$= 0.73\angle 0.41$$

so that

$$y_p(k) = \text{Re}\,\{0.73\, e^{j0.41}\, e^{j6k}\}$$

$$= 0.73 \cdot 2^k \cos(6k + 23.25^o)$$

which is equivalent to the previous result. The computational advantage of the method of the present section should be readily apparent in the example.

## 4.8 Simultaneous Sets of Difference Equations

As with other differential equations we are often confronted with simultaneous sets of difference equations. To facilitate the solution of these types of equations we introduce the shift operator $S^n$ where

$$S^n y(k) \triangleq y(k+n) \tag{43}$$

Similar to ordinary differential equations, the shift operator has the property that

$$S^n S^{-n} y(k) = y(k) \tag{44}$$

The $n$-th order difference equation in (5) can be written as

$$(S^n + a_{n-1} S^{n-1} + \ldots + a_0)\, y(k) = f(k) \tag{45}$$

In terms of the shift operator, a simultaneous set such as

$$2y(k+1) - y(k) + 4x(k+2) = 2^k \tag{46}$$

$$y(k+1) + y(k) + 2x(k+1) - 3x(k) = 0$$

may be written as

$$(2S - 1)\, y(k) + (4S^2)\, x(k) = 2^k \tag{47}$$

$$(S + 1)\, y(k) + (2S - 3)\, x(k) = 0$$

To solve this set we must eliminate $x(k)$ and determine a difference equation involving only $y(k)$ and vice-versa. To do so we multiply or operate on the first equation in (47) with $-(2S-3)$ and the second equation with $(4S^2)$ resulting in

$$-(2S - 3)(2S - 1)\, y(k) - (2S - 3)(4S^2)\, x(k) = -(2S - 3) \cdot 2^k \tag{48}$$

$$= -2 \cdot 2^{(k+1)} + 3 \cdot 2^k$$

$$(4S^2)(S+1)\,y(k) + (4S^2)(2S-3)\,x(k) = 0$$

Noting that (2S - 3) $(4S^2)$ x(k) = $(4S^2)$ (2S - 3) x(k) we subtract to yield

$$[(4S^2)(S+1) - (2S-3)(2S-1)]\,y(k) = -2{\cdot}2^{(k+1)} + 3{\cdot}2^k \qquad (49)$$

or

$$[4S^3 + 8S - 3]\,y(k) = 4y(k+3) + 8y(k+1) - 3y(k)$$
$$= -2{\cdot}2^{(k+1)} + 3{\cdot}2^k \qquad (50)$$

and have obtained a single difference equation involving only $y(k)$.

This technique can be placed in a matrix format by writing a general set of $n$ difference equations in the $n$ unknowns, $y_1(k), y_2(k), \ldots, y_n(k)$ as:

$$\underbrace{\begin{bmatrix} L_{11}(S) & L_{12}(S) & \ldots & L_{1n}(S) \\ L_{21}(S) & L_{22}(S) & \ldots & L_{2n}(S) \\ & & & \\ L_{n1}(S) & \ldots & \ldots & L_{nn}(S) \end{bmatrix}}_{\mathbf{L}(S)} \begin{bmatrix} y_1(k) \\ y_2(k) \\ \vdots \\ y_n(k) \end{bmatrix} = \begin{bmatrix} f_1(k) \\ f_2(k) \\ \vdots \\ f_n(k) \end{bmatrix} \qquad (51)$$

where each $L_{ij}$(S) is a polynomial in the shift operator, $S$. Since the shift operator obeys the rules of algebra we may use methods of Chapter 1 to solve (51). For example we may obtain

$$\Delta(S) \begin{bmatrix} y_1(k) \\ y_2(k) \\ \vdots \\ y_n(k) \end{bmatrix} = \Delta(S)\,\mathbf{L}^{-1}(S) \begin{bmatrix} f_1(k) \\ f_2(k) \\ \vdots \\ f_n(k) \end{bmatrix} \qquad (52)$$

where $\Delta(S) = |\mathbf{L}(S)|$. From (52) we may obtain $n$ difference equations each involving only one variable, $y_i(k)$.

*Example 4-6*

Determine difference equations involving $y_1(k)$ and $y_2(k)$ from the following set:

$$(S+3)\,y_1(k) + 2\,y_2(k) = 2^k$$

$$(S+2)\,y_1(k) + (S+1)\,y_2(k) = 1$$

*Solution*

In matrix notation

$$\mathbf{L}(S) = \begin{bmatrix} (S+3) & 2 \\ (S+2) & (S+1) \end{bmatrix}$$

with determinant

$$\Delta(S) = S^2 + 2S - 1$$

The inverse is

$$\mathbf{L}^{-1}(S) = \frac{1}{\Delta(S)} \begin{bmatrix} (S+1) & -2 \\ -(S+2) & (S+3) \end{bmatrix}$$

so that

$$(S^2 + 2S - 1) \begin{bmatrix} y_1(k) \\ y_2(k) \end{bmatrix} = \begin{bmatrix} (S+1) & -2 \\ -(S+2) & (S+3) \end{bmatrix} \begin{bmatrix} 2^k \\ 1 \end{bmatrix}$$

$$= \begin{bmatrix} 2^{(k+1)} + 2^k & -2 \\ -2^{k+1} - 2 \cdot 2^k & +1+3 \end{bmatrix}$$

$$= \begin{bmatrix} 2^{(k+1)} + 2^k & -2 \\ -2^{(k+1)} - 2 \cdot 2^k & 4 \end{bmatrix}$$

from which we obtain

$$y_1(k+2) + 2y_1(k+1) - y_1(k) = 2^{(k+1)} + 2^k - 2$$

$$y_2(k+2) + 2y_2(k+1) - y_2(k) = -2^{(k+1)} - 2 \cdot 2^k + 4$$

As with ordinary differential equations the undetermined constants in the homogeneous solutions for $y_1(k), y_2(k), \ldots, y_n(k)$ are interrelated by the difference equation set and the initial conditions on $y_1(k), y_2(k), \ldots, y_n(k)$ are not all independently specifiable. The number of independently specifiable initial conditions is equal to the degree of $\Delta(S)$.

## Solved Problems

*4.1* Twenty one 1-ohm resistors are connected to a 10V dc voltage source in a ladder arrangement as shown in Fig. P4.1. Determine the current $I_{21}$ in the last resistor, $R_{21}$.

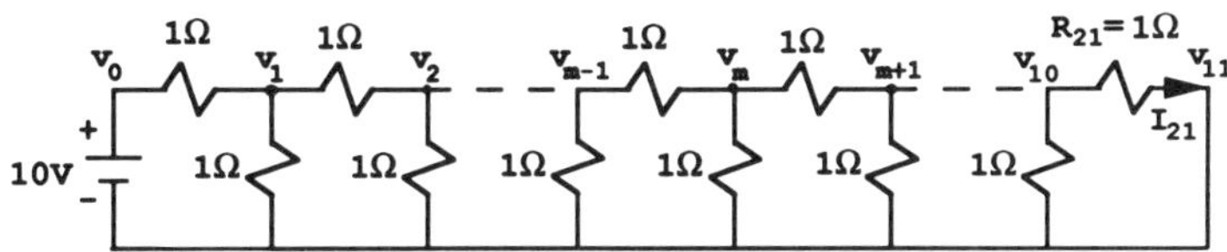


Figure P4.1

*Solution*

Writing node voltage equations for a typical section consisting of three nodes

$$-V_{m-1} + 3\,V_m - V_{m+1} = 0$$

for $m = 1, 2, \ldots$. In the form of a difference equation we may write this as

$$-V(k+2) + 3V(k+1) - V(k) = 0$$

where $k = m - 1$ and $k = 0,1,2, \ldots$. The characteristic equation is

$$-r^2 + 3r - 1 = 0$$

or

$$r^2 - 3r + 1 = 0$$

with roots

$$r_1, r_2 = \frac{3}{2} \pm \frac{\sqrt{5}}{2}$$

so that $r_1 = 2.62$, $r_2 = .38$. Thus

$$V(k) = V_h(k) = C_1 \cdot 2.62^k + C_2 \cdot .38^k$$

The condition at the left end is $V_0 = 10$ or $V(0) = 10$. At the right end, $V_{11} = 0$ or $V(11) = 0$. Thus

$$V(0) = 10 = C_1 + C_2$$

$$V(11) = 0 = C_1 \cdot 2.62^{11} + C_2 \cdot .38^{11}$$

or

$$\begin{bmatrix} 1 & 1 \\ 2.62^{11} & .38^{11} \end{bmatrix} \begin{bmatrix} C_1 \\ C_2 \end{bmatrix} = \begin{bmatrix} 10 \\ 0 \end{bmatrix}$$

Solving yields

$$C_1 = \frac{10 \cdot .38^{11}}{.38^{11} - 2.62^{11}}$$

$$C_2 = \frac{-10 \cdot 2.62^{11}}{.38^{11} - 2.62^{11}}$$

Thus

$$V_{10} = V(10) = \frac{10 \cdot 0.38^{11} \cdot 2.62^{10} - 10 \cdot 2.62^{11} \cdot .38^{10}}{.38^{11} - 2.62^{11}}$$

$$= 5.65 \times 10^{-4}$$

A large number of digits need to be carried in this computation to avoid round-off error.

*4.2* Once every cycle $k$ in a chemical process, $u(k)$ liters of chemical $A$ and $100–u(k)$ liters of chemical $B$ are added to 900 liters of a mixture in a large vat where $0 \le u(k) \le 100$ and $k = 1, 2, 3, \ldots$. The vat contents are thoroughly mixed and 100 liters of mixture are drawn off. Let $y(k)$ be the fractional concentration of chemical $A$ in the mixture drawn off, that is, 1000 $y(k)$ is the amount of chemical $A$ in the vat. Develop a difference equation for $y(k)$.

*Solution*

We relate the total amount of chemical $A$ at cycle $k$ to the amount of chemical at cycle $k-1$ plus any inputs. Thus

$$1000\, y(k) = 900\, y(k-1) + u(k)$$

or

$$y(k) - 0.9\, y(k-1) = 0.001\, u(k)$$

$$k=1,2,3,\ldots$$

*4.3* Consider the discrete time system shown in Fig. P4.3. Determine the output sequence $y(k)$ if the input sequence is $u(k) = 3^k$ for $k \ge 0$ and $u(k)=0$ for $k \le 0$.

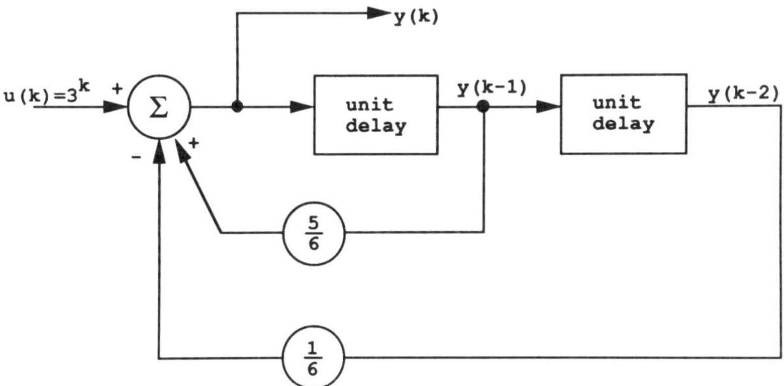


Figure P4.3

*Solution*

From the block diagram we obtain

$$y(k) = \frac{5}{6}\, y(k-1) - \frac{1}{6}\, y(k-2) + 3^k$$

for $k \geq 0$ or

$$y(k+2) - \frac{5}{6} y(k+1) + \frac{1}{6}\, y(k) = 3^{k+2}$$

The characteristic equation is

$$r^2 - \frac{5}{6}\, r + \frac{1}{6} = 0$$

with roots $r_1 = {}^1\!/\!_2$ and $r_2 = \frac{1}{3}$. The homogeneous solution is

$$y_h(k) = C_1\,({}^1\!/\!_2)^k + C_2\,(\frac{1}{3})^k$$

The particular solution is found by assuming

$$y_p(k) = K\, 3^k$$

Substitution yields

$$(3^2 - \frac{5}{6} \cdot 3 + \frac{1}{6})\, K\, 3^k = 9 \cdot 3^k$$

or

$$K = \frac{9}{3^2 - \frac{5}{6} \cdot 3 + \frac{1}{6}}$$

$$= 1.35$$

Thus

$$y(k) = C_1 (^1/_2)^k + C_2 (\frac{1}{3})^k + 1.35 \cdot 3^k$$

CHAPTER

# 5

# NUMERICAL METHODS

## 5.1 General Remarks

In the preceding chapters, we have obtained exact or analytical solutions to various types of equations. Situations arise in numerous engineering problems where exact solutions to equations governing various physical phenomena are not forthcoming. In such cases, we resort to approximate solutions in the form of numerical end results. With the availability of hand-held electronic calculators or access to digital computers, in many problems it may be advantageous to obtain numerical solutions. Of course, in many other cases solutions by numerical methods are the only possibility. Comparing analytical solutions to numerical solutions, for most engineering problems (especially boundary value problems) it may be said that exact solutions are feasible for approximate (or idealized) models, whereas approximate methods are applicable to exact models.

Because most computations are carried out on a digital computer, it is advantageous to use a procedure or *algorithm* suitable for execution on the computer. Algorithms involving numerical methods contain *errors* (resulting from arithmetic operations). A *round-off error* results from the use of a limited number of significant figures in the computation. *Truncation error* arises because of approximating the solution of an equation by a numerical method. For instance, a truncation error will result from the neglect of higher order terms in the Taylor series expansion in the finite-difference approximation to a partial differential equation, as we shall see later.

In addition to errors, in an algorithm we should consider also the following points. (a) *Condition* of the problem such that if the data of the problem are perturbed slightly, the answer to the problem must not be affected substantially. (b) *Stability* of algorithm such that the algorithm produces a sufficiently accurate result for a given problem. (c) *Efficiency* of algorithm, *computation time* and *convergence* to the solution.

*Example 5.1*

An example of ill-conditioning is the determination of the roots of the equation

$$\prod_{n=1}^{20} (x-n) = 0 \tag{1}$$

Obviously, the roots of the equation are the integers 1, 2, ..., 20. However, if the coefficient of $x^{19}$ is changed from $-210$ to $(210+2^{-23})$ the exact roots which were 16 and 17 become a pair of complex conjugate roots: $16.73 \ldots \pm j2.81 \ldots$.

Numerical methods have a wide range of applications to mathematical problems. However, in the following we will restrict ourselves to the determination of the roots of a polynomial and to numerical solutions of the types of equations presented in Chapters 1 through 4.

## 5.2 Roots of a Polynomial

There are a number of numerical methods which are applicable to solving the polynomial equation:

$$f(x) = x^n + a_1 x^{n-1} + a_2 x^{n-2} + \ldots + a_n = 0 \tag{1}$$

(with $a$'s real) and thus finding its roots. Certain methods, such as the method of false position or the incremental-search method, yield only the real roots. These are not considered here. Of the various methods available for finding both real and complex roots, Graeffe's root-squaring technique, quotient-difference method (or Rutihauser's QD algorithm), and iterative factorization of polynomials (or Bairstow's method) are the most general. Only the last mentioned method is considered here. Before proceeding with this method, we should keep the following points in mind.

(a) There are $n$ roots of (1). The roots may be real, distinct, coincident, or complex.
(b) If $n$ is an odd integer, at least one of the roots will be real.
(c) The number of positive roots is equal to the number of sign changes of the coefficients or is less than this number by an even integer.
(d) The rule of the number of negative roots is also (c) above.
(e) Complex roots occur in conjugate pairs.

### 5.2.1 Iterative Factorization of Polynomials: Bairstow's Method

Bairstow's method is particularly suited for solution on a digital computer, but for convergence more than one trial may be necessary. The method is an algorithm for finding quadratic factors iteratively for the polynomial (1). Thus, if a quadratic factor is

$$\phi(x) = x^2 + ux + v \tag{2}$$

the roots are

$$r_{1,2} = {}^1\!/_2 \left[-u \pm \sqrt{u^2 - 4v}\,\right] \tag{3}$$

From (3) it is clear when the roots are real and distinct ($u^2 > 4v$), coincident ($u^2 = 4v$), or a pair of complex conjugates ($u^2 < 4v$). Having obtained two roots, (3), from the first quadratic factor, (2), two more roots are obtained by repeating the iteration process. The procedure is continued until all the roots are found.

In numerical computation, a starting value of the $uv$-pair is chosen. By iteration, we converge to the correct values of $u$ and $v$. However, the iteration process may not converge from the chosen initial values. Suitable starting valves are: $u = a_{n-1}/a_{n-2}$ and $v = a_n/a_{n-2}$, although $u = v = 0$ is also used as a starting point.

To formulate the iteration equations, we express (1) as

$$(x^2 + ux + v)(x^{n-2} + b_1x^{n-3} + b_2x^{n-4} + \ldots + b_{n-3}x + b_{n-2}) + R = 0 \quad (4)$$

where $R$ = remainder. Comparing (1) and (4) we find that $b$'s and $R$ are given by

$$\begin{aligned} b_0 &= 1 \\ b_1 &= a_1 - u \\ b_2 &= a_2 - b_1u - v \\ b_3 &= a_3 - b_2u - b_1v \\ &\vdots \\ b_n &= a_n - b_{n-1}u - b_{n-2}v \end{aligned} \quad (5)$$

and

$$R = (x + u)b_{n-1} + b_n \quad (6)$$

Clearly, if $u$ and $v$ are such that (2) is a factor, then $R = 0$ and $b_n = b_{n-1} = 0$. Thus, the objective of the method is to reduce the remainder to close to zero, implying that $b_{n-1} = 0$ and $b_n = 0$. But, $b$'s are functions of $u$ and $v$. So, we let $\Delta u$ and $\Delta v$ be respective increments to $u$ and $v$ to make $R = 0$, and we expand $b_n$ and $b_{n-1}$ in Taylor series.

$$b_n(u + \Delta u, v + \Delta v) = 0 \quad b_n = \frac{\partial b_n}{\partial u}\Delta u + \frac{\partial b_n}{\partial v}\Delta v \quad (7)$$

$$b_{n-1}(u + \Delta u, v + \Delta v) = 0 \quad b_{n-1} = \frac{\partial b_{n-1}}{\partial u}\Delta u + \frac{\partial b_{n-1}}{\partial u}\Delta v$$

Taking partial derivatives of (5) with respect to $u$, and expressing in terms of $c$'s we obtain

$$\begin{aligned} \frac{\partial b_1}{\partial u} &= -1 = -c_0 \\ \frac{\partial b_2}{\partial u} &= u - b_1 = -c_1 \\ \frac{\partial b_3}{\partial u} &= -b_2 + c_1u + v = -c_2 \\ &\vdots \end{aligned} \quad (8)$$

$$\frac{\partial b_n}{\partial u} = -b_{n-1} + c_{n-2}u + c_{n-3}v = -c_{n-1}$$

A similar set of equations can be written for partial derivatives of (5) with respect to $v$. From (8) it follows that

$$c_j = b_j - c_{j-1}u - c_{j-2}v \qquad (j = 2,3,\dots, n-1)$$
$$c_0 = 1 \tag{9}$$
$$c_1 = b_1 - u$$

Also, from (8) we may write (7) as

$$b_n = c_{n-1}\,\Delta u + c_{n-2}\,\Delta v$$
$$b_{n-1} = c_{n-2}\Delta u + c_{n-3}\,\Delta v \tag{10}$$

which may be solved for $\Delta u$ and $\Delta v$:

$$\Delta u = \frac{\begin{vmatrix} b_n & c_{n-2} \\ b_{n-1} & c_{n-3} \end{vmatrix}}{\begin{vmatrix} c_{n-1} & c_{n-2} \\ c_{n-2} & c_{n-3} \end{vmatrix}} \quad \text{and} \quad \Delta v = \frac{\begin{vmatrix} c_{n-1} & b_n \\ c_{n-2} & b_{n-1} \end{vmatrix}}{\begin{vmatrix} c_{n-1} & c_{n-2} \\ c_{n-2} & c_{n-3} \end{vmatrix}} \tag{11}$$

The various steps in the procedure are now summarized.

Step 1. Choose $u_1 = v_1 = 0$.
Step 2. Determine $b_0, b_1, b_2, \dots, b_n$ from (5).
Step 3. Calculate $c_0, c_1, c_2, \dots, c_n$ from (9).
Step 4. Evaluate $\Delta u$ and $\Delta v$ from (11).
Step 5. Increment $u$ and $v$ by $\Delta u$ and $\Delta v$ to get

$$u_{i+1} = u_i + \Delta u_i \tag{12}$$
$$v_{i+1} = v_i + \Delta v_i$$

Step 6. Return to Step 2 and repeat the procedure until $\Delta u$ and $\Delta v$ are within a predetermined small value $\varepsilon$ such that

$$|\Delta u_{i+1}| + |\Delta v_{i+1}| < \varepsilon \tag{13}$$

Step 7. Get the first two roots from (3).
Step 8. Obtain additional roots from the reduced polynomial formulated from the final values of the $b$'s.

*Example 5.1*

Find the roots of the following polynomial by Bairstow's method.

$$f(x) = x^5 - 15x^4 + 85x^3 - 225x^2 + 274x - 120$$

It can be readily verified that it is extremely cumbersome, if not impossible, to find the roots on a hand-held electronic calculator. By trial and error it may be seen that the roots of the polynomical are: 1,2,3,4 and 5.

A computer program (in FORTRAN) to obtain the roots of the above polynomial by Bairstow's method is given in Fig. 5.1, which also shows the end results. However, before solving for the roots on a computer, it would be helpful to consider the following observations.

```
C       ROOTS OF A POLYNOMIAL BY BAIRSTOWS METHOD
        DIMENSION A(30),B(30),C(30)
        WRITE(6,2)
    2   FORMAT('1',9X,'REAL PART',3X,'IMAGINARY PART',3X,'ITERATIONS'//)
        READ(5,3) UI,VI,EPSI,N
    3   FORMAT(3F10.0,I2)
        READ(5,4) (A(I),I=1,N)
    4   FORMAT(10F8.0)
C       SEE IF N=0,N=1,OR N IS GREATER THAN 1
   40   IF(N-1) 100,5,7
    5   P=-A(1)
        Q=0.0
        IT=1
        WRITE(6,6)N,P,Q,IT
    6   FORMAT('','X(',I2,') =',2X,F8.4,6X,F8.4,10X,I3)
        GO TO 100
C       SEE IF N=2 OR N IS GREATER THAN 2
    7   IF(N.EQ.2) GO TO 8
        GO TO 13
    8   U=A(1)
        V=A(2)
        IT=1
    9   P=-U/2.0
        RAD=U**2-4.0*V
C       CHECK THE SIGN OF U**2-4.0*V
        IF(RAD.GT.0.0)GO TO 12
        RAD=-RAD
        Q=SQRT(RAD)/2.0
        WRITE(6,6)N,P,Q,IT
        N=N-1
        Q=-Q
   90   WRITE(6,6)N,P,Q,IT
   10   N=N-1
C       CHECK TO SEE IF N IS GREATER THAN 0
        IF(N.LE.0) GO TO 100
        DO 11 I=1,N
   11   A(I)=B(I)
        GO TO 40
   12   Q=SQRT(RAD)/2.0
        W=P
        Z=Q
        P=P+Q
        Q=0.0
        WRITE(6,6)N,P,Q,IT
        N=N-1
        P=W-Z
        GO TO 90
   13   U=UI
        V=VI
        IT=1
C       CALCULATE THE B VALUES
   50   B(1)=A(1)-U
        B(2)=A(2)-B(1)*U-V
        DO 14 K=3,N
   14   B(K)=A(K)-B(K-1)*U-B(K-2)*V
C       CALCULATE THE C VALUES
        C(1)=B(1)-U
        C(2)=B(2)-C(1)*U-V
        M=N-1
        DO 15 K=3,M
```

```
Page 2      File: BAIRS    FORTRAN  A1  University of Kentucky

   15  C(K)=B(K)-C(K-1)*U-C(K-2)*V
 C     CALCULATE DELU AND DELV
       IF(N.GT.3) GO TO 17
       DENOM=C(N-1)-C(N-2)**2
       IF(DENOM.EQ.0.0) GO TO 30
       DELU=(B(N)-B(N-1)*C(N-2))/DENOM
   16  DELV=(C(N-1)*B(N-1)-C(N-2)*B(N))/DENOM
       GO TO 18
   17  DENOM=C(N-1)*C(N-3)-C(N-2)**2
       IF(DENOM.EQ.0.0) GO TO 30
       DELU=(B(N)*C(N-3)-B(N-1)*C(N-2))/DENOM
       GO TO 16
 C     CALCULATE NEW U AND V VALUES
   18  U=U+DELU
       V=V+DELV
       SUM=ABS(DELU)+ABS(DELV)
 C     STORE THE FIRST SUM CACULATED
       IF(IT.EQ.1) GO TO 19
       GO TO 20
   19  STORE=SUM
       IF(IT.EQ.200) GO TO 26
   21  IF(SUM.LE.EPSI) GO TO 9
       IF(IT.EQ.100) GO TO 23
   22  IT=IT+1
       GO TO 50
   23  WRITE(6,24)
   24  FORMAT(' ',10X,'CONVERGENCE IS SLOW')
       WRITE(6,25) U,V,DELU,DELV
   25  FORMAT(' ','U=',E14.7,3X,'V=',E14.7,3X,'DELU=',E14.7,3X,'DELV=',
      *E14.7)
       GO TO 22
   26  WRITE(6,27)
   27  FORMAT(' ',10X,'ITERATING STOPPED AFTER 200 ITERATIONS')
       WRITE(6,25) U,V,DELU,DELV
       GO TO 100
 C     SEE SUM AFTER 50 ITERATIONS EXCEEDS THE FIRST SUM STORED
   28  IF(SUM.LT.STORE) GO TO 21
       WRITE(6,29)
   29  FORMAT(' ',10X,'DIVERGENCE OCCURING')
       WRITE(6,25)U,V,DELU,DELV
       GO TO 100
   30  WRITE(6,31)
   31  FORMAT(' ',10X,'DENOM IS ZERO')
       GO TO 100
  100  STOP
       END
  $ENTRY
             REAL PART    IMAGINARY PART    ITERATIONS

  X(5) =      5.0000         0.0000            17
  X(4) =      1.0000         0.0000            17
  X(3) =      3.0000         0.0000            10
  X(2) =      2.0000         0.0000            10
  X(1) =      4.0000         0.0000             1
```

Figure 5.1

The choice of the initial values of $u$ and $v$, and the value of $\varepsilon$, of (13), is critical in obtaining accuracy and convergence. For instance, starting with $u=v=0$ and $\varepsilon=0$, convergence could not be achieved after 200 iterations. It is interesting to note that even 10,000 iterations did not yield convergence indicating that the number of iterations is not a critical factor. As a recommended set of initial values of $u$ and $v$ (to accelerate convergence) we may start with

$$u = \frac{a_n}{a_{n-2}} = \frac{274}{-225} = -1.218 \text{ and } v = \frac{a_{n-1}}{a_{n-2}} = \frac{-120}{-225} = 0.533.$$

With these values, the effect of $\varepsilon$ on accuracy and convergence is illustrated in Fig. 5.2(a). On the other hand, Fig. 5.2(b) shows the effect of $\varepsilon$ if we had started with $u=v=0$ (but $\varepsilon \neq 0$). The conclusion is that the combination $(u,v,\varepsilon)$ should be chosen judiciously.

## 5.3 Linear Simultaneous Algebraic Equations

In Chapter 1 we discussed Cramer's rule and Gauss elimination method of solving a set of linear simultaneous algebraic equations. When the number of equations is large (say 5 or more) it is convenient to carry out the computations on a digital computer. Furthermore, in many engineering problems, it is often desirable just to obtain numerical end results. In the following, we will first consider the Gauss elimination procedure numerically. Next, we will present the method of matrix decomposition and factorization (Crout's method). This will be followed by the matrix inversion method.

| X(5) | X(4) | X(3) | X(2) | X(1) | EPSI |
|---|---|---|---|---|---|
| 5.0000(15) | 1.0000(15) | 3.0000(14) | 2.0000(14) | 4.0000(01) | 0.000005 |
| 5.0000(15) | 1.0000(15) | 3.0000(14) | 2.0000(14) | 4.0000(01) | 0.00001 |
| 5.0000(15) | 1.0000(15) | 3.0000(08) | 2.0000(08) | 4.0000(01) | 0.00005 |
| 5.0000(15) | 1.0000(15) | 3.0000(08) | 2.0000(08) | 4.0000(01) | 0.00010 |
| 5.0000(15) | 1.0000(15) | 3.0000(07) | 2.0000(07) | 3.9998(01) | 0.00050 |
| 5.0000(14) | 1.0000(14) | 3.0010(07) | 1.9996(07) | 3.9989(01) | 0.00100 |
| 5.0000(14) | 1.0000(14) | 3.0010(07) | 1.9996(07) | 3.9989(01) | 0.00500 |
| 5.0000(14) | 1.0000(14) | 3.0010(07) | 1.9996(07) | 3.9989(01) | 0.01000 |
| 5.0003(13) | 1.0000(13) | 3.0420(07) | 1.9829(07) | 3.9614(01) | 0.05000 |
| 5.0116(12) | 1.0019(12) | 3.4358(08) | 1.9376(08) | 3.5583(01) | 0.10000 |
| 5.0116(12) | 1.0019(12) | 3.3952(06) | 1.9313(06) | 3.7276(01) | 0.50000 |
| 5.2490(10) | 1.0942(10) | 3.5166(37) | 3.5166(37) | 1.2687(01) | 1.00000 |

(a)

| X(5) | X(4) | X(3) | X(2) | X(1) | EPSI |
|---|---|---|---|---|---|
| 5.0000(19) | 1.0000(19) | 3.0001(15) | 2.0000(15) | 3.9999(01) | 0.00001 |
| 5.0000(12) | 1.0000(12) | 3.0001(08) | 2.0000(08) | 3.9999(01) | 0.00005 |
| 5.0000(10) | 1.0000(10) | 3.0002(08) | 1.9999(08) | 3.9999(01) | 0.00010 |
| 5.0000(09) | 1.0000(09) | 2.9998(08) | 2.0002(08) | 4.0001(01) | 0.00050 |
| 5.0000(09) | 1.0000(09) | 2.9998(08) | 2.0002(08) | 4.0001(01) | 0.00100 |
| 5.0000(09) | 1.0000(09) | 2.9998(07) | 2.0002(07) | 3.9996(01) | 0.00500 |
| 5.0000(09) | 1.0000(09) | 2.9998(07) | 2.0002(07) | 3.9996(01) | 0.01000 |
| 5.0000(08) | 0.9999(08) | 2.9829(07) | 2.0166(07) | 4.0056(01) | 0.05000 |
| 5.0000(08) | 0.9999(08) | 2.9829(07) | 2.0166(07) | 4.0056(01) | 0.10000 |
| 4.9997(07) | 0.9946(07) | 2.7603(06) | 2.2234(06) | 4.0989(01) | 0.50000 |
| 4.9894(06) | 0.9468(06) | 2.4159(04) | 2.4159(04) | 4.8818(01) | 1.00000 |

(b)

Figure 5.2

### 5.3.1 Gauss Elimination Method

In addition to the material of Chapter 1, in order to use Gauss elimination method, we must define two concepts.

(a) *Augmented matrix* - In a set of $n$ equations involving $n$ unknowns, if $[a]$ is the coefficient matrix and $[b]$ is a column matrix denoting the constants (on the right-hand sides of the equations), then the augmented matrix is formed by adding the column $[b]$ in the position of the $[a]$ matrix given in (14):

$$[A] = \begin{bmatrix} a_{11} & a_{12} & \ldots & a_{1n} & b_1 \\ a_{21} & a_{22} & \ldots & a_{2n} & b_2 \\ a_{n1} & a_{n2} & \ldots & a_{nn} & b_n \end{bmatrix} \tag{14}$$

(b) *Pivot equation* - It is the equation used to eliminate the unknowns in the equations which follow the pivot equation. In this equation the coefficient of the unknown to be eliminated from subsequent equations is the *pivot coefficient.*

We now recall from Chapter 1 that Gauss elimination requires the original set of equations (2) to be reduced to an equivalent triangular form (3). In the process, the first step was to divide the first equation in (2) by the coefficient of $x_1$. Thus, the elements of the reduced augmented matrix $[A']$ are:

$$a'_{ij} = a_{ij} - \frac{a_{ik}}{a_{kk}} (a_{kj}) \qquad \begin{matrix} k \le j \le m \\ (k+1) \le i \le n \end{matrix} \tag{15}$$

where $a_{ij}$ = element of the $[A]$ - matrix in the $i$th row and $j$th column
$a'_{ij}$ = element of the reduced matrix $[A']$ and the matrix is $n$ by $m$.

To represent an interative procedure for numerical computations we re-write (15) as:

$$a_{ij}^{p} = a_{ij}^{p-1} - \frac{a_{kj}^{p-1}}{a_{kk}^{p-1}} (a_{ik}^{p-1}) \qquad \begin{matrix} (k+1) \le j \le m \\ (k+1) \le i \le n \end{matrix} \tag{16}$$

where the superscripts correspond to the successive reduced matrix computations.

The back substitution, discussed in Chapter 1, is given in a general form as:

$$x_n = \frac{a_{nm}}{a_{nn}}$$

$$x_i = \frac{1}{a_{ii}} [a_{im} - \sum_{j=i+1}^{n} a_{ij} \, x_j] \quad i = (n-1),\ (n-2),\ \ldots,\ 1 \tag{17}$$

Before proceeding with computations, it must be checked that a pivot row has all nonzero elements. If there is a zero, or very small, pivot element, the row having the zero (or small) element is interchanged by another row (not having a zero or small element). If the pivot element is much smaller than other elements in the augmented matrix, then the accuracy of the solution is reduced. Thus, for accuracy, in each reduction step the pivot row should have the largest pivot element. This is called *partial pivoting*. In *complete pivoting* rows are interchanged as well columns.

*Example 5.2*

Solve the following set of simultaneous equations by Gauss elimination method

$$x + y + z = 6$$

$$2x - 3y + 2z = 2$$

$$3x + y - z = 2$$

Although the above set of equations can be solved very conveniently by elimination, a computer program for Gauss elimination method is given in Fig. 5.3.

### 5.3.2 Crout's Method

Crout's method involves matrix decomposition and factorization, requires less computer time than Gauss elimination method, and is especially suited for structural analysis. Suppose we wish to solve (2) of Chapter 1, rewritten in matrix form as

$$[A][X] = [B] \tag{18}$$

where $[A]$ = coefficient matrix, and $[X]$ and $[B]$ are column matrices of unknowns and constants respectively. Let (18) be reduced to an equivalent triangular form such that

$$\begin{bmatrix} 1 & u_{12} & u_{13} & \dots & u_{1n} \\ 0 & 1 & u_{23} & \dots & u_{2n} \\ . & . & . & & . \\ . & . & . & & . \\ . & . & . & & . \\ 0 & 0 & 0 & & 1 \end{bmatrix} \begin{bmatrix} x_1 \\ x_2 \\ . \\ . \\ . \\ x_n \end{bmatrix} = \begin{bmatrix} c_1 \\ c_2 \\ . \\ . \\ . \\ c_n \end{bmatrix} \tag{19}$$

```
C       SOLUTION OF SIMULTANEOUS EQUATIONS BY GAUSSIAN ELIMINATION
        DIMENSION A(15,16),X(15)
        READ(5,3) N
   3    FORMAT(I3)
        M=N+1
        L=N-1
   5    READ(5,4) ((A(I,J),J=1,M),I=1,N)
   4    FORMAT(8F10.0)
        DO 12 K=1,L
        JJ=K
        BIG=ABS(A(K,K))
        KP1=K+1
C
C
C       SEARCH FOR LARGEST POSSIBLE PIVOT ELEMENT
        DO 7 I=KP1,N
        AB=ABS(A(I,K))
        IF(BIG-AB) 6,7,7
   6    BIG=AB
        JJ=I
   7    CONTINUE
C
C
C       DECISION ON NECESSITY OF ROW INTERCHANGE
        IF(JJ-K) 8,10,8
C
C
C       ROW INTERCHANGE
   8    DO 9 J=K,M
        TEMP=A(JJ,J)
        A(JJ,J)=A(K,J)
   9    A(K,J)=TEMP
C
C
C       CALCULATION OF ELEMENTS OF NEW MATRIX
  10    DO 11 I=KP1,N
        QUOT=A(I,K)/A(K,K)
        DO 11 J=KP1,M
  11    A(I,J)=A(I,J)-QUOT*A(K,J)
        DO 12 I=KP1,N
  12    A(I,K)=0.0
C
C
C       FIRST STEP IN BACK-SUBSTITUTION PROCESS
        X(N)=A(N,M)/A(N,N)
C
C
C       REMAINDER OF BACK-SUBSTITUTION
        DO 14 NN=1,L
        SUM=0.0
        I=N-NN
        IP1=I+1
        DO 13 J=IP1,N
  13    SUM=SUM+A(I,J)*X(J)
  14    X(I)=(A(I,M)-SUM)/A(I,I)
        WRITE(6,1)
   1    FORMAT('1','X(1) THROUGH X(N)'/)
        DO 15 I=1,N
  15    WRITE(6,2) X(I)
   2    FORMAT('',E14.8)
        STOP
        END
   $ENTRY
 X(1) THROUGH X(N)
 0.99999960E00
 0.20000000E01
 0.30000000E01
```

Figure 5.3

Or,

$$[U]\,[X] = [C] \tag{20}$$

where $[U]$ is known as the upper triangular matrix. Now, suppose there exists a lower triangular matrix $[L]$ which has the following property:

$$[L]\,[U]\,[X] - [L]\,[C] = [A]\,[X] - [B] \tag{21}$$

Then

$$[L]\,[U] = [A] \tag{22}$$

and

$$[L]\,[C] = [B] \tag{23}$$

Knowing $[L]$ and $[U]$, $[X]$ can be found by back substitution. The elements of $[L]$ and the augmented $[U]$ matrices are given by:

$$\ell_{i1} = a_{i1}, \qquad i = 1,2,\ldots,n \tag{24a}$$

$$u_{ij} = \frac{a_{ij}}{a_{ii}} \qquad j = 1,2,\ldots,(n+1) \tag{24b}$$

$$\ell_{ij} = a_{ij} - \sum_{k=1}^{j-1} \ell_{ik}\, u_{kj}, \quad j = 2,3,\ldots,n \tag{24c}$$

$$i = j,(j+1),\ldots,n$$

$$u_{ij} = \frac{1}{\ell_{ii}}\,[a_{ij} - \sum_{k=1}^{i-1} \ell_{ik}\, u_{kj}], \quad i = 2,3,\ldots,n \tag{24d}$$

$$j = (i+1), (i+2), \ldots (n+1)$$

The computational steps are as follows:

Step 1.

Apply (24a) and (24b) over the ranges of $i$ and $j$.

Step 2.

Apply (24c) with $j$=2 and $i$ ranging from $j$ to $n$.

Step 3.

Apply (24d) with $i = 2$ and $j$ ranging from $(i+1)$ to $(n+1)$

Step 4.

Apply (24c) with $j$=3 and $i$ ranging from $j$ to $n$.

Step 5.

Apply (24d) with $i = 3$ and $j$ ranging from $(i+1)$ to $(n+1)$.

And so on.

The general expressions for the back-substitution are:

$$x_n = u_{n,n+1} \tag{25a}$$

$$i=(n-1),(n-2),\ldots,1$$

$$x_i = u_{i,n+1} - \sum_{j=i+1}^{n} u_{ij} x_j \tag{25b}$$

### 5.3.3 Matrix Inversion

We may solve (2) of Chapter 1, or (18), by matrix inversion such that

$$[X] = [A]^{-1}[B] = [D]\,[B] \tag{26}$$

The elements of $[D]$ or $[A]^{-1}$ are given by

$$d_{i-1,j-1} = a_{ij} - \frac{a_{ij}a_{i1}}{a_{11}}, \quad 1 < i \le n \tag{27a}$$

$$1 < j \le m$$

$$a_{11} \ne 0$$

$$d_{n,j-1} = \frac{a_{1j}}{a_{11}}, \quad 1 < j \le m \tag{27b}$$

$$a_{11} \ne 0$$

where the reduction process is continued until $n$ columns remain.

## 5.4 Ordinary Differential Equations

There is a large variety of numerical methods available for solving ordinary differential equations. These methods include: direct numerical integration, Euler's, Runge-Kutta, Milne's, Hamming's, and predictor-corrector methods. For our purposes we shall divide the problems relating to differential equations into the class of (a) initial-value problems, or (b) boundary-value problems. And, in the following we shall consider only a few of the numerical methods of solving ordinary differential equations.

For initial-value problems, which are discussed first, we shall discuss only Runge-Kutta and predictor-corrector methods. We define an *initial-value problem* as that in which the values of the dependent variables and/or its derivatives are given at the point at which integration begins. In contrast, in a *boundary-value problem*, conditions are specified at both ends of the interval of integration. We have considered several boundary-value problems in Chapter 3.

### 5.4.1 Runge-Kutta Methods

A Runge-Kutta method uses a recurrence formula of the form

$$y_{i+1} = y_i + \sum_{j=1}^{n} a_j k_j \tag{28}$$

to determine successive values of y of the differential equation

$$y' = \frac{dy}{dx} = f(x,y) \tag{29}$$

In (28),

$$k_j = hf\left(x_i + p_{j-1}h,\, y_i + \sum_{m=1}^{j-1} q_{j-1,m}\, k_m\right)$$

$$p_0 = 0$$

$$\text{and } \sum_{m=1}^{j-1} q_{j-1,m}\, k_m = 0 \text{ for } j = 1$$

Let us consider the development of a second-order Runge-Kutta method. Thus, for $n = 2$, (28) becomes

$$y_{i+1} = y_i + a_1 k_1 + a_2 k_2 \tag{30}$$

with

$$k_1 = hf(x_i, y_i)$$

$$k_2 = hf(x_i + p_1 h,\, y_i + q_{1,1} k_1)$$

We must determine $a_1, a_2, p_1$, and $q_{1,1}$ such that (30) is satisfied. This is accomplished by making (30) equivalent to a truncated Taylor-series expansion of $y$ about $x_i$. Expanding $y_{i+1}$ about $x_i$ we get

$$y_{i+1} = y_i + hy'_i + \frac{h^2}{2!} y''_i + \ldots \tag{31}$$

where the primes denote differential coefficients.
From (29) we have

$$y'_i = f(x_i, y_i) \tag{32}$$

From the chain rule of differentiation we get, if $y' = f(x, y)$,

$$dy' = \frac{\partial f}{\partial x} dx + \frac{\partial f}{\partial y} dy$$

Or,

$$\frac{dy'}{dx} = \frac{\partial f}{\partial x} + \frac{\partial f}{\partial y}\frac{dy}{dx} \tag{33}$$

Therefore, (32) and (33) yield

$$y''_i = \frac{\partial f}{\partial x}(x_i, y_i) + \frac{\partial f}{\partial y}(x_i, y_i)\, f(x_i, y_i) \tag{34}$$

and (31) - (34) give

$$y_{i+1} = y_i + hf(x_i, y_i) + \frac{h^2}{2!}\left[\frac{\partial f}{\partial x}(x_i, y_i) + \frac{\partial f}{\partial y}(x_i, y_i) f(x_i, y_i)\right] + \ldots \tag{35}$$

Observe that $k_2$ is a function of two variables and can be expanded in a Taylor series about $(x_i, y_i)$ to yield

$$k_2 = h\,[f(x_i, y_i) + p_1 h \frac{\partial f}{\partial x}(x_i, y_i) + q_{1,1} k_1 \frac{\partial f}{\partial y}(x_i, y_i)] \tag{36}$$

Consequently, from (30), (35) and (36) we get

$$y_{i+1} = y_i + a_1 hf(x_i, y_i) + a_2 h\, f(x_i, y_i) + a_2 h^2\,[p_1 \frac{\partial f}{\partial x}(x_i, y_i) + q_{1,1}\frac{\partial f}{\partial y}(x_i, y_i) f(x_i, y_i)] \tag{37}$$

Comparing coefficients of (35) and (37) yields

$$\begin{aligned} a_1 + a_2 &= 1 \\ a_2 p_1 &= 1/2 \\ a_2 q_{1,1} &= 1/2 \end{aligned} \tag{38}$$

In order to determine four unknowns from three equations, (38), we arbitrarily assign a value to one unknown. Let $a_1 = 1/2$ in (38). Then, $a_2 = 1/2$, $p_1 = q_{1,1} = 1$, and (30) becomes

$$y_{i+1} = y_i + {}^1\!/\!_2\,(k_1+k_2) \tag{39}$$

with

$$k_1 = h\,f\,(x_i, y_i)$$

$$k_2 = hf\,(x_i+h, y_i+k_1)$$

Equation (39) can be used to solve (29), and the solution will have a per-step truncation error of the order of $h^3$.

Without going into the details of derivations, algorithms for the third- and fourth-order Runge-Kutta methods are as follows.
Third-order Runge-Kutta method-

$$y_{i+1} = y_i + \frac{1}{6}\,(k_1+4k_2+k_3) \tag{40}$$

with

$$k_1 = hf\,(x_i, y_i)$$

$$k_2 = hf\,(x_i + \frac{h}{2}, y_i+\frac{k_1}{2})$$

$$k_3 = hf\,(x_i + h, y_i - k_1 + 2k_2)$$

Fourth-order (or classic) Runge-Kutta method -

$$y_{i+1} = y_i + \frac{1}{6}\,(k_1 + 2k_2 + 2k_3 + k_4) \tag{41}$$

with

$$k_1 = hf\,(x_i, y_i)$$

$$k_2 = hf\,(x_i + \frac{h}{2}, y_i + \frac{k_1}{2})$$

$$k_3 = hf\,(x_i + \frac{h}{2}, y_i + \frac{k_2}{2})$$

$$k_4 = hf\,(x_i + h, y_i + k_3)$$

Per-step truncation errors in the third- and fourth-order methods are, respectively, of the order of $h^4$ and $h^5$.

Runge-Kutta methods have the advantage that they are self-starting and are easy to program.

## 5.4.2 Solutions of Simultaneous Differential Equations

In the preceding section, we have presented Runge-Kutta methods of solving first order ordinary differential equations. In many engineering applications, the governing equations are not necessarily first-order equations. However, it can be shown that an $n$th-order differential equation can be expressed as a set of $n$ first-order simultaneous equations, which may be solved by a Runge-Kutta method.

We first consider the representation of an $n$th order equation by $n$ first-order simultaneous equations. The procedure is illustrated by the following example.

*Example 5.3*

Express the following second-order equation by two first-order equations.

$$a_1\ddot{x} + a_2\dot{x} + a_3 x = f(t) \tag{1}$$

where the "dots" denote differential coefficients (with respect to $t$).

We write (1) as

$$\ddot{x} = -\frac{a_2}{a_1}\dot{x} - \frac{a_3}{a_1}x + \frac{1}{a_1}f(t) \tag{2}$$

and let

$$y_1 = x \tag{3a}$$

$$y_2 = \dot{x} = \dot{y}_1 \tag{3b}$$

$$\dot{y}_2 = \ddot{x}$$

From (2) and (3) we get

$$\dot{y}_2 = -\frac{a_2}{a_1}y_2 - \frac{a_3}{a_1}y_1 + \frac{1}{a_1}f(t)$$

$$\dot{y}_1 = y_2$$

which are the two first order equations.

Returning now to the solutions of simultaneous equations, let us consider two general simultaneous first-order differential equations:

$$\frac{du}{dx} = \psi[x, u(x), v(x)] \tag{42a}$$

$$\frac{dv}{dx} = \phi[x, u(x), v(x)] \tag{42b}$$

with $u(x_0) = u_0$ and $v(x_0) = v_0$

Using (41), for (42a) we may write

$$u_{i+1} = u_i + \frac{1}{6}(k_1+2k_2+2k_3+k_4) \tag{43}$$

with

$$k_1 = hf(x_i, u_i, v_i)$$

$$k_2 = hf(x_i + \frac{h}{2}, u_i + \frac{k_1}{2}, v_i + \frac{q_1}{2})$$

$$k_3 = hf(x_i + \frac{h}{2}, u_i + \frac{k_2}{2}, v_i + \frac{q_2}{2})$$

$$k_4 = hf(x_i+h, u_i+k_3, v_i+q_3)$$

A similar set of equations can be written for (42b).

We solve (42 a and b) beginning with the initial values $(u_0, v_0)$ to obtain the initial values of $\psi$ and $\phi$. Next, $k_1$ and $q_1$ are obtained by multiplying the initial values of $\psi$ and $\phi$ by $h$, the step increment. With $k$, and $q$, known, $k_2$ and $q_2$, followed by $k_3$ and $q_3$, and $k_4$ and $q_4$ are determined. Then, from (43) $u_{i+1}$ is evaluated, and a similar recurrence formula gives $v_{i+1}$. These new $(u, v)$ - values are used as beginning values for repeating the above procedure, until the desired range of integration is covered.

### 5.4.3 Predictor - Corrector Methods

There are several so-called predictor-corrector methods of numerically solving a first-order ordinary differential equation. Here, we will consider only Milne's method. Let us solve

$$y' = f(x, y)$$

with $y(0) = y_0$. The predictor-corrector method consists of obtaining approximate values of $y$ by a predictor equation, and subsequently correcting these values of $y$ by the iterative use of a corrector equation. The method is not self-starting, but the starting values of $y$ may be obtained by (say) the fourth-order Runge-Kutta method. The details of the procedure are as follows.

Establish the values of $y_1$,$y_2$ and $y_3$ by means of (41). Determine the corresponding $y'_1$, $y'_2$, and $y'_3$ from (44). Then use the *predictor equation*:

$$P(y_{i+1}) = y_{i-3} + \frac{4}{3} h \, (2y'_i - y'_{i-1} + 2y'_{i-2}) \tag{45}$$

Substitute the predicted values in (44) such that

$$P(y'_{i+1}) = f\,[x_{i+1} + P(y_{i+1})] \tag{46}$$

Next, use $P(y'_{i+1})$ from (46) in the *corrector equation*

$$C(y_{i+1}) = y_{i-1} + \frac{h}{3} \, [y'_{i-1} + 4y'_i + P \, (y'_{i+1})] \tag{47}$$

to obtain a corrected value of $y_{i+1}$. This corrected value of $y_{i+1}$ is substituted into the differential equation

$$C(y'_{i+1}) = f\,[x_{i+1} + C(y_{i+1})] \tag{48}$$

to get a corrected $y'_{i+1}$. This value is next used in the iterating corrector equation

$$C(y_{i+1}) = y_{i-1} + \frac{h}{3} \, [y'_{i-1} + 4y'_i + C(y'_{i+1})] \tag{49}$$

to obtain a further improved $y_{i+1}$, which is substituted back in (48) to get a further improved value of $y'_{i+1}$, and so on. The iterations continue until successive values of $y_{i+1}$ differ by a predetermined small number. In (45)-(49) $P$ and $C$ respectively correspond to predictor and corrector equations. The above method is more efficient (in terms of computer time) compared to the Runge-Kutta method.

## 5.5 Partial Differential Equations

A number of numerical methods exist which can be used to solve partial differential equations. For instance, Laplace's equation in a given region can be solved numerically by the finite-element, the finite-difference method, or the method of moments. In the following, the finite-difference method will be applied to static field problems.

### 5.5.1 Finite-Difference Method

In obtaining a solution by the finite-difference method, the partial differential equation (to be solved) is replaced by a system of algebraic equations. The solution to the system of algebraic equations consists of values at discrete points in the region of interest, whereas the solution to the corresponding partial differential equation gives the values everywhere in the region. Because the system of algebraic equations is generally large, the solution is preferably obtained with the aid of a digital computer. Furthermore, to speed up the process of obtaining the solutions, certain special techniques are used (as will be mentioned shortly). Also, the computation is terminated as soon as a predetermined accuracy is achieved. Thus, in applying numerical methods we must pay attention to the convergence, error, and accuracy of solutions. Some of these aspects will be considered later in the subsection. But first we observe that a set of finite-difference equations is valid for a given distribution of discrete points in the field. The various standard distributions of these discrete points, called *nodes*, are as follows.

In principle, any form of distribution on nodes may be used, but to simplify the problem, a regular distribution of nodes is almost invariably chosen. A network of square meshes is shown in Fig. 5.4(a). The arrangement of nodes in the square mesh, such as nodes 01234 (redrawn in Fig. 5.4b), is known as the *computational molecule* or *characteristic star*. It is a symmetrical star since the distances between the outer nodes and the center node, the mesh length, are the same. We choose the mesh length $h$ to be small compared with the dimensions of the field boundaries. For a symmetrical star, we now proceed to obtain the finite-difference equation corresponding to Laplace's equation.

We consider Laplace's equation in two dimensions,

$$\frac{\partial^2 V}{\partial x^2} + \frac{\partial^2 V}{\partial y^2} = 0 \tag{50}$$

where $V$ is the potential in a given region shown in Fig. 5.5. This region is divided into a square mesh of side $h$. If the unknown $V$'s at the indicated nodes are as shown and if $h$ is sufficiently small, then for a point $a$ midway between the nodes 0 and 1 we have

$$\frac{\partial V}{\partial x}\bigg|_a \cong \frac{1}{h}(V_1 - V_0) \tag{51}$$

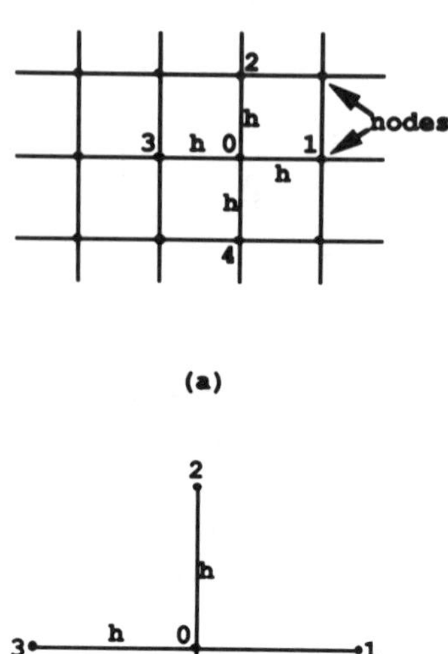


Figure 5.4

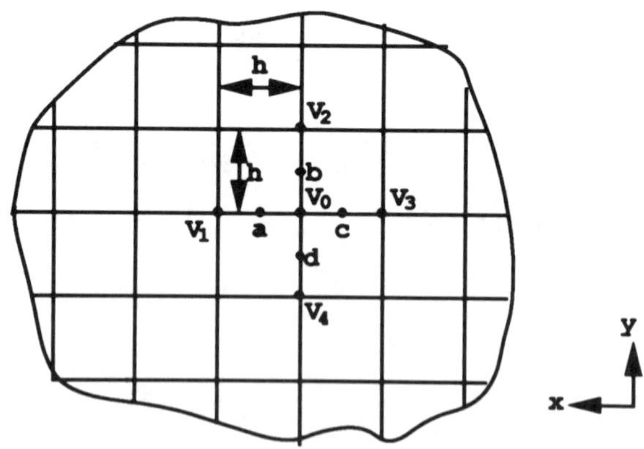


Figure 5.5

And, similararaly, for the point $c$ we may write

$$\frac{\partial V}{\partial x}\Big|_c \cong \frac{1}{h}(V_0 - V_3) \tag{52}$$

so that at the point 0, from (51) and (52) we obtain

$$\frac{\partial^2 V}{\partial x^2}\Big|_0 \cong \frac{1}{h}\left[\frac{\partial V}{\partial x}\Big|_a - \frac{\partial V}{\partial x}\Big|_c\right] \cong \frac{1}{h^2}[(V_1 - V_0) - (V_0 - V_3)] \tag{53}$$

Similarly, if we consider the potentials along the $y$ axis, we have

$$\frac{\partial^2 V}{\partial y^2}\Big|_0 \cong \frac{1}{h^2}[(V_2 - V_0) - (V_0 - V_4)] \tag{54}$$

Equations (53) and (54), when substituted in (118), yield

$$V_0 \cong {}^1\!/_4\,(V_1 + V_2 + V_3 + V_4) \tag{55}$$

Thus, from (55), $V_0$ is found from the potential values at the adjacent corners of every square in the mesh. In practice, the computations are carried out in a number of ways, the objectives being that (55) be satisfied at every node and that the computations take a minimum time. Although hand computation is possible for a small number of nodes, it is best to carry out the computations on a digital computer. The procedure for hand computation is illustrated by the following two examples, which show that the finite-difference method involves the repeated modification of the potentials at each node until (55) is satisfied at the nodes within a predetermined degree of accuracy.

As mentioned previously, the iterative method is based on using (55) at every node and on using it repeatedly at the nodes until (55) is satisfied within reasonably accuracy. The iterative procedure for determining the potentials at given nodes is illustrated in the following example.

*Example 5.4*

Compute the potentials at the nodes $a, b, c, d, e$, and $f$ shown in Fig. 5.6, which also shows the potentials at the boundaries.

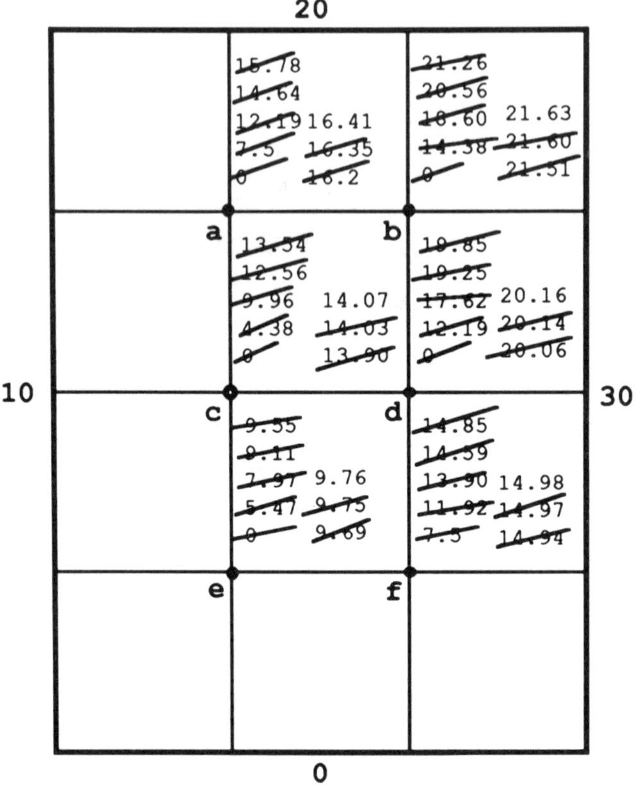


Figure 5.6

*Solution*

The potential function $V$ satisfies Laplace's equation. First we inscribe the square mesh, as shown, and assume that all the specified nodes are at zero potential except the potential at node $f$. Then we apply (55) at each node in succession. For instance, for the first cycle of computation, $V_f = \frac{1}{4}(30 + 0 + 0 + 0) = 7.5$, $V_a = \frac{1}{4}(10 + 20 + 0 + 0) = 7.5$, and so on. For the next cycle of computation, we use the previously calculated potentials to compute new potentials at the various nodes. Thus, for example, at the end of the second cycle we have $V_f = \frac{1}{4}(5.47 + 12.19 + 30 + 0) = 11.92$, and so on. This process is continued for eight cycles of computations — or until none of the potentials at the six nodes changes by a predetermined value (say less than 0.5 percent) from its value computed for the preceding cycle (i.e., from its value last crossed out).

In Example 5.4 we obtained the potentials at each node by successive approximation (or repeated modification) of the potentials at each node until (55) was satisfied to a given degree of accuracy. However, we had no measure of the accuracy of the computed potential at a given node during a particular cycle. The degree of accuracy is determined by the residuals at the nodes. Referring to (55), if for a particular node 0 we have

$$V_1 + V_2 + V_3 + V_4 - 4V_0 = R_0 \tag{56}$$

then $R_0$ is the residual at node 0. Obviously, for the correct values of the potentials $V_0, V_1, \ldots, V_4$, the residual is zero. The *relaxation method* aims at reducing the residuals to zero, or to a predetermined small value, at all the nodes by suitable variations in $V_0$'s. To achieve an accuracy of 1 to 5 percent, it is generally recommended that the individual residuals eventually be reduced to 0.1 percent of the mean of all the node potentials. This procedure of reducing the residual at one node at a time is known as *point relaxation* and is a very convenient method for hand computation. The method is illustrated in the following example.

*Example 5.5*

Solve the problem of Example 5.4 by the point-relaxation method.

*Solution*

As in the last example, first we lay out the square mesh as shown in Fig. 5.7 and assume that all the specified nodes are at zero potential.

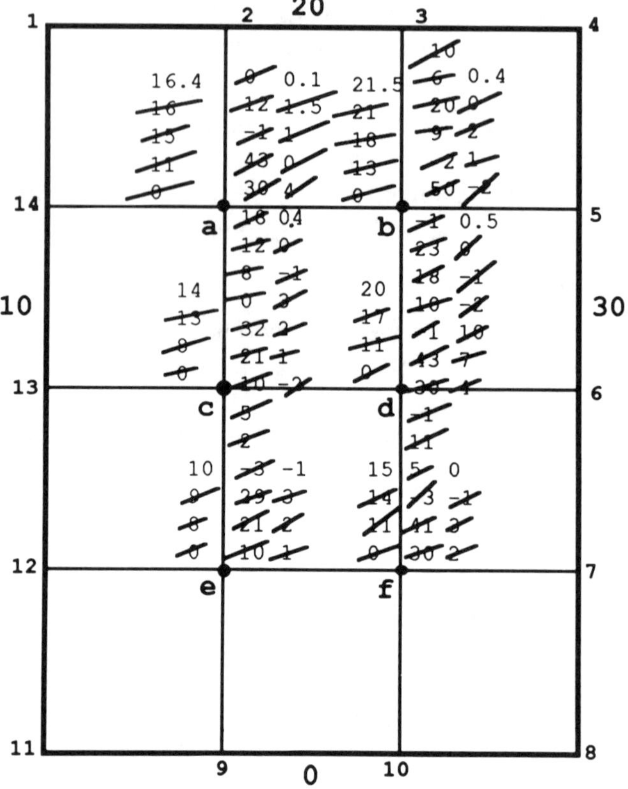


Figure 5.7

Conventionally, we record the potentials at the top left of each node and write the values of the residuals at the top right of the node. While going through each cycle of computation, we write down the new values of potentials and residuals and cross out the old ones, as shown in Fig. 5.7. Notice that we have attempted to reduce the largest residual occurring at each stage approximately to zero. For the first cycle, the node $b$ has the largest residual, 50. The mean of the surrounding node potentials $V_a$, $V_d$, $V_5$ and $V_3$ is $50/4 \cong 13$. Therefore, the potential at node $b$ is changed to 13; from (56) the resulting residual at this node becomes 2, and this changes the residuals at nodes $a, b$, and $d$. Next we observe that the largest residual of 43 occurs at node $a$. The mean of the surrounding node potentials is 11, so we change the potential of node $a$ to 11; this change brings about a change in the residuals at nodes $a, b$, and $c$. We look for the node with the largest residual again and adjust the potential at that node to make the residual close to zero. This process is continued until the residuals at all the nodes are reduced to a predetermined minimum.

The last two examples illustrate the two common methods of hand computation of potentials at specified nodes. Clearly, hand computation is cumbersome and becomes very time-consuming if a large number of nodes in involved. For such cases, solutions are generally obtained by a digital computer. Reconsidering (55), we may write down the following set of equations for the nodes $a, b, c, d, e$, and $f$ of Fig. 5.7. Notice that the nodes at the boundaries have been designated as nodes 1 through 14. Thus, we have

$$\begin{aligned}
-4V_a + V_b + V_c &= -V_2 - V_{14} \\
V_a - 4V_b + V_d &= -V_3 - V_5 \\
V_a - 4V_c + V_d + V_e &= -V_{13} \\
V_b + V_c - 4V_d + V_f &= -V_6 \\
V_c - 4V_e + V_f &= -V_{10} - V_{12} \\
V_d + V_e - 4V_f &= -V_7 - V_9
\end{aligned} \tag{57}$$

In matrix notation, we may rewrite (57) as

$$\underset{[\mathrm{A}]}{\begin{bmatrix} -4 & 1 & 1 & 0 & 0 & 0 \\ 1 & -4 & 0 & 1 & 0 & 0 \\ 1 & 0 & -4 & 1 & 1 & 0 \\ 0 & 1 & 1 & -4 & 0 & 1 \\ 0 & 0 & 1 & 0 & -4 & 1 \\ 0 & 0 & 0 & 1 & 1 & -4 \end{bmatrix}} \underset{[\mathrm{V}]}{\begin{bmatrix} V_a \\ V_b \\ V_c \\ V_d \\ V_e \\ V_f \end{bmatrix}} = \underset{[\mathrm{B}]}{\begin{bmatrix} -V_2 - V_{14} \\ -V_3 - V_5 \\ -V_{13} \\ -V_6 \\ -V_{10} - V_{12} \\ -V_7 - V_9 \end{bmatrix}} \tag{58}$$

or,

$$[\mathrm{A}]\,[\mathrm{V}] = [\mathrm{B}] \tag{59}$$

Premultiplying both sides of (59) by $[\mathrm{A}]^{-1}$ yields the desired solution:

$$[\mathrm{V}] = [\mathrm{A}]^{-1}[\mathrm{B}] \tag{60}$$

Various standard numerical methods exist to find $[\mathrm{A}]^{-1}$. Notice that $[\mathrm{A}]$ is a *band matrix* as marked in (58), and for such a case special numerical techniques may be used to obtain a rapid convergence to the final solution.

We close this section by briefly considering the errors affecting the accuracy of numerical solutions. The three common errors are: (1) *round-off error*, (2) *truncation error*, and (3) *computational error*. Round-off error occurs because we can use only a limited number of significant figures in the computations. The truncation error is introduced by expressing a partial differential equation as a difference equation, where certain higher-order terms are neglected (as in the Taylor series expansion of the partial derivatives in Laplace's equation). The computational error is caused by limiting the number of iterations to a finite number in obtaining the solution via iterative methods.

*Example 5.6*

Obtain the solution to the boundary-value problem of Example 5.4.

*Solution*

In this case, the equation to be solved is the matrix equation (59), in which the elements of [**B**] are (-30, -50, -10, -30, -10, -30). The system of equations in (59) may be solved by using a computer, which gives these results:

| | *Equation (60)* | *Iterative method (Fig. 5.6)* |
|---|---|---|
| $V_a$ | 16.44 | 16.41 |
| $V_b$ | 21.66 | 21.63 |
| $V_c$ | 14.10 | 14.07 |
| $V_d$ | 20.19 | 20.16 |
| $V_e$ | 9.77 | 9.76 |
| $V_f$ | 14.99 | 14.98 |

APPENDIX

# A

# COMPLEX NUMBERS

## A.1 Definitions and Basic Rules

A complex number $z$ is defined by

$$z = x + jy \tag{A.1}$$

where $x$ and $y$ are real numbers and $j$ is defined by

$$j^2 = -1$$

or,

$$j = \sqrt{-1} \tag{A.2}$$

In (A.1) $x \equiv$ real part of $z$, or $Re\ (z)$; and $y \equiv$ imaginary part of $z$, or $Im\ (z)$.

Complex numbers obey the laws of algebra, with the understanding that $j^2 = -1$, and $z = 0$ only when $x = y = 0$. Thus, for addition (and subtraction) we have

$$z_1 \pm z_2 = (x_1 \pm x_2) + j\,(y_1 \pm y_2) \tag{A.3}$$

And for multiplication we get

$$z_1 z_2 = (x_1 x_2 - y_1 y_2) + j\,(x_1 y_2 - x_2 y_1) \tag{A.4}$$

Finally, the quotient of two numbers can be written as:

$$\frac{z_1}{z_2} = \frac{x_1 + jy_1}{x_2 + jy_2} = \frac{(x_1 + jy_1)(x_2 - jy_2)}{(x_2 + jy_2)(x_2 - jy_2)} = \frac{x_1 x_2 + y_1 y_2}{x_2^2 + y_2^2} \tag{A.5}$$

$$+ j\,\frac{x_2 y_1 - x_1 y_2}{x_2^2 + y_2^2}$$

From (A.3) - (A.5) we see that as end results of basic arithmetic operations, we can separate real and imaginary parts; that is, we can always write the result as $z = x + jy$.

Two complex numbers differing only in the sign of their imaginary parts are called complex conjugate pairs. Thus, if $z = x + jy$ then its complex conjugate is $z^* = x - jy$.

Two complex numbers, $z_1 = x_1 + jy_1$ and $z_2 = x_2 + jy_2$ are equal if and only if $x_1 = x_2$ and $y_1 = y_2$.

## A.2 Geometric and Trigonometric Representations

A point P is shown in Fig. A.1. This point, having the coordinates $(x, y)$ or $(r, \theta)$, may be said to correspond to a complex number

$$z = x + jy \tag{A.6}$$

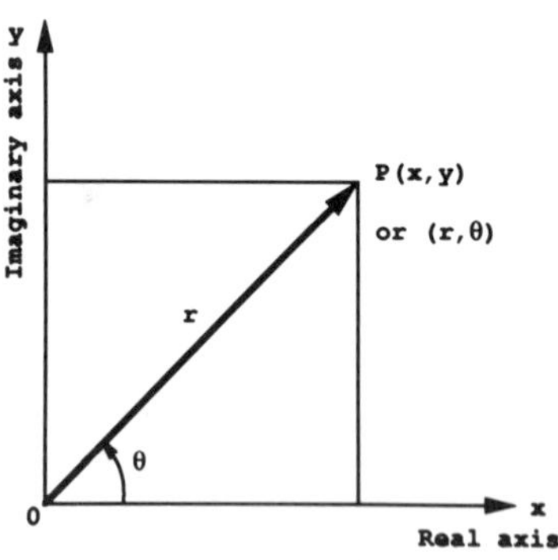


Figure A.1

where the $x$-axis ≡ the axis of reals, or simply real axis and the $y$-axis ≡ imaginary axis.

In polar coordinates, since $x = r\cos\theta$ and $y = r\sin\theta$, we may write (A.6) as

$$z = r(\cos\theta + j\sin\theta) \qquad \text{(A.7)}$$

In (A.7), $r$ is always taken positive, and is called the modulus or absolute value of $z$. Thus

$$r = |z| = \sqrt{x^2 + y^2} \qquad \text{(A.8)}$$

and

$$\tan\theta = \frac{y}{x}$$

From Maclaurin expansions of sine, cosine, and exponential, it may be shown that

$$\cos\theta \pm j\sin\theta = e^{\pm j\theta} \qquad \text{(A.9)}$$

and

$$(\cos\theta + j\sin\theta)^n = \cos n\theta + j\sin n\theta \qquad \text{(A.10)}$$

Combining (A.7) and (A.9) we get

$$z = r\,e^{j\theta} = |z|e^{j\theta} \qquad \text{(A.11)}$$

Equation (A.11) is the polar form of $z$ whereas (A.6) is the Cartesian form. The polar form is especially useful in handling multiplication and division of complex numbers. Hence

$$z_1 z_2 = r_1 r_2\, e^{j(\theta_1 + \theta_2)} \qquad \text{(A.12)}$$

and

$$\frac{z_1}{z_2} = \frac{r_1}{r_2}\, e^{j(\theta_1 - \theta_2)} \qquad \text{(A.13)}$$

APPENDIX

# B

# FOURIER SERIES AND FOURIER INTEGRAL

## B.1 Fourier Series

A single-valued periodic function $f(x)$ having finite number of discontinuities, and finite number of maxima and minima in one period, and for which the integral $\int_{-\pi}^{\pi} |f(x)|\, dx$ is finite ($2\pi$ being the period), can be expanded in a Fourier series as follows:

$$f(x) = \tfrac{1}{2} a_0 + \sum_{n=1}^{\infty} (a_n \cos nx + b_n \sin nx) \tag{B.1}$$

where the coefficients are given by

$$a_n = \frac{1}{\pi} \int_{-\pi}^{\pi} f(x) \cos nx \, dx \qquad n = 0,1,2,... \tag{B.2}$$

$$b_n = \frac{1}{\pi} \int_{-\pi}^{\pi} f(x) \sin nx \, dx \qquad n = 1,2,3,... \tag{B.3}$$

## B.2 Even and Odd Series

If $f(x)$ is such that $f(-x) = f(x)$ then $f(x)$ is an even function. Its Fourier expansion contains no sin terms and is simply

$$f(x) = \tfrac{1}{2} a_0 + \sum_{n=1}^{\infty} a_n \cos nx \tag{B.4}$$

where

$$a_n = \frac{2}{\pi} \int_0^{\pi} f(x) \cos nx \, dx \tag{B.5}$$

On the other hand, if $f(-x) = -f(x)$ then $f(x)$ is an odd function. It contains only *sin* terms, and its Fourier expansion is

$$f(x) = \sum_{n=1}^{\infty} b_n \sin nx \tag{B.6}$$

where

$$b_n = \frac{2}{\pi} \int_0^{\pi} f(x) \sin nx \, dx \tag{B.7}$$

*Example B.1*

Obtain the Fourier series for the rectangular train of pulses shown in Fig. B.1.

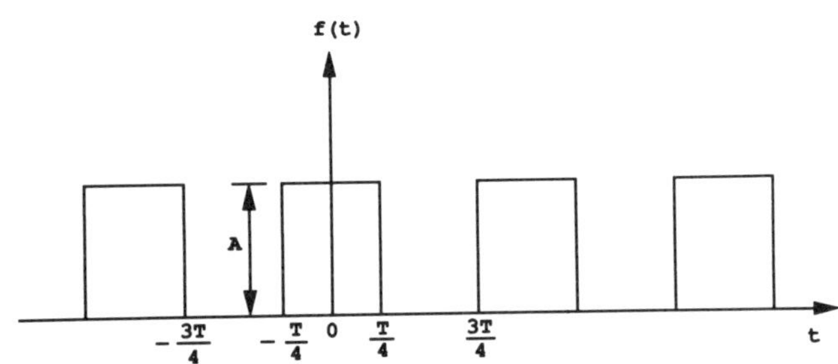


Figure B.1

From Fig. B.1 we have:

$$f(t) = 0 \qquad -T/2 < t < -T/4$$
$$= A \qquad -T/4 < t < T/4$$

$$= 0 \qquad T/4 < t < T/2$$

We observe that $f(t)$ is an even function. Thus, (B.4) is the form of expansion. We define the period $T$ such that $T = 2\pi/\omega$ where $\omega$ is called the angular frequency. From (B.5) we get

$$a_n = \frac{2}{T} \int_{-T/2}^{T/2} f(t) \cos n\omega t \, dt$$

$$= \frac{2}{T} \int_{-T/4}^{T/4} A \cos n\omega t \, dt$$

$$= \frac{2A}{n\pi} \sin \frac{n\pi}{2} \qquad n \text{ odd}$$

Hence,

$$f(t) = \frac{A}{2} + \frac{2A}{\pi} \left(\cos \omega t - \frac{1}{3} \cos 3\omega t + \frac{1}{5} \cos 5\omega t - \ldots\right)$$

$$= \frac{A}{2} + \frac{2A}{\pi} \sum_{k=1}^{\infty} (-1)^{k-1} \frac{1}{(2k-1)} \cos (2k-1)\omega t.$$

Notice from the above example that the period of the function is $T$ (rather than $2\pi$, as in Sections B.1 and B.2). Thus, the Fourier coefficients of a function having a period $T$ are:

$$a_n = \frac{2}{T} \int_{-T/2}^{T/2} f(t) \cos \frac{2\pi n}{T} t \, dt \tag{B.8}$$

and

$$b_n = \frac{2}{T} \int_{-T/2}^{T/2} f(t) \sin \frac{2\pi n}{T} t \, dt \tag{B.9}$$

## B.3 Fourier Integral

Reconsider (B.1). Replace $f(x)$ by $f(t)$ of a period $T$. Let $A_n = \frac{T}{2} a_n$, $B_n = \frac{T}{2} b_n$ and $\omega_n = \frac{2\pi n}{T}$. Then (B.1) becomes

$$f(t) = \frac{A_0}{T} + \frac{2}{T} \sum_{n=1}^{\infty} (A_n \cos \omega_n t + B_n \sin \omega_n t) \tag{B.10}$$

Define $\Delta\omega_n = \omega_{n+1} - \omega_n = \frac{2\pi}{T}$. Then (B.10) becomes

$$f(t) = \frac{A_0}{T} + \frac{1}{\pi} \sum_{n=1}^{\infty} (A_n \cos \omega_n t \, \Delta\omega_n + B_n \sin \omega_n t \, \Delta\omega_n) \qquad \text{(B.11)}$$

Now, $\omega_0 = 0$, $\sin \omega_0 t = 0$ and $\cos \omega_0 t = 1$, so that

$$\frac{A_0}{T} = \frac{A_0}{2\pi} \cos \omega_0 t \, \Delta\omega_0 \qquad \text{(B.12)}$$

From (B.11) and (B.12) we obtain

$$f(t) = -\frac{A_0}{T} + \frac{1}{\pi} \sum_{n=0}^{\infty} (A_n \cos \omega_n t \, \Delta\omega_n + B_n \sin \omega_n t \, \Delta\omega_n) \qquad \text{(B.13)}$$

If we let $T \to \infty$, then $-A_0/R \to 0$ and the sums in (B.13) approach integrals, and (B.13) becomes

$$f(t) = \frac{1}{\pi} \int_0^{\infty} [A(\omega) \cos \omega t + B(\omega) \sin \omega t] \, d\omega \qquad \text{(B.14)}$$

where

$$A(\omega) = \int_{-\infty}^{\infty} f(t) \cos \omega t \, dt \qquad \text{(B.15)}$$

and

$$B(\omega) = \int_{-\infty}^{\infty} f(t) \sin \omega t \, dt \qquad \text{(B.16)}$$

Equation (B.14) is called the Fourier integral and (B.15) and (B.16) are known as Fourier transforms. If $f(t)$ = even, then $B(\omega) \equiv 0$; and if $f(t)$ = odd then $A(\omega) \equiv 0$.

APPENDIX

# C

# VECTOR OPERATIONS AND VECTOR IDENTITIES

## C.1 Vector Operations

Rectangular coordinates $(x, y, z)$.

$$\nabla\Phi = \mathbf{a}_x \frac{\partial\phi}{\partial x} + \mathbf{a}_y \frac{\partial\phi}{\partial y} + \mathbf{a}_z \frac{\partial\phi}{\partial z}$$

$$\nabla \cdot \mathbf{D} = \frac{\partial D_x}{\partial x} + \frac{\partial D_y}{\partial y} + \frac{\partial D_z}{\partial z}$$

$$\nabla \times \mathbf{H} = \mathbf{a}_x \left[\frac{\partial H_z}{\partial y} - \frac{\partial H_y}{\partial z}\right] + \mathbf{a}_y \left[\frac{\partial H_x}{\partial z} - \frac{\partial H_z}{\partial x}\right] + \mathbf{a}_z \left[\frac{\partial H_y}{\partial x} - \frac{\partial H_x}{\partial y}\right]$$

$$\nabla^2\Phi = \frac{\partial^2\Phi}{\partial x^2} + \frac{\partial^2\Phi}{\partial y^2} + \frac{\partial^2\Phi}{\partial z^2}$$

$$\nabla^2\mathbf{A} = \mathbf{a}_x \nabla^2 A_x + \mathbf{a}_y \nabla^2 A_y + \mathbf{a}_z \nabla^2 A_z$$

Cylindrical coordinates $(r,\phi,z)$.

$$\nabla\Phi = \mathbf{a}_r\frac{\partial\Phi}{\partial r} + \mathbf{a}_\phi\frac{1}{r}\frac{\partial\Phi}{\partial\phi} + \mathbf{a}_z\frac{\partial\Phi}{\partial z}$$

$$\nabla\cdot\mathbf{D} = \frac{1}{r}\frac{\partial}{\partial r}(rD_r) + \frac{1}{r}\frac{\partial D_\phi}{\partial\phi} + \frac{\partial D_z}{\partial z}$$

$$\nabla\times\mathbf{H} = \mathbf{a}_r\left[\frac{1}{r}\frac{\partial H_z}{\partial\phi} - \frac{\partial H_\phi}{\partial z}\right] + \mathbf{a}_\phi\left[\frac{\partial H_r}{\partial z} - \frac{\partial H_z}{\partial r}\right] + \mathbf{a}_z\left[\frac{1}{r}\frac{\partial(rH_\phi)}{\partial r} - \frac{1}{r}\frac{\partial H_r}{\partial\phi}\right]$$

$$\nabla^2\Phi = \frac{1}{r}\frac{\partial}{\partial r}\left(r\frac{\partial\Phi}{\partial r}\right) + \frac{1}{r^2}\frac{\partial^2\Phi}{\partial\phi^2} + \frac{\partial^2\Phi}{\partial z^2}$$

$$\nabla^2\mathbf{A} = \mathbf{a}_r\left(\nabla^2A_r - \frac{2}{r^2}\frac{\partial A_\phi}{\partial\phi} - \frac{A_r}{r^2}\right) + \mathbf{a}_\phi\left(\nabla^2A_\phi + \frac{2}{r^2}\frac{\partial A_r}{\partial\phi} - \frac{A_\phi}{r^2}\right) + \mathbf{a}_z(\nabla^2A_z)$$

---

Spherical coordinates $(r,\theta,\phi)$.

$$\nabla\Phi = \mathbf{a}_r\frac{\partial\Phi}{\partial r} + \mathbf{a}_\theta\frac{1}{r}\frac{\partial\Phi}{\partial\theta} + \frac{\mathbf{a}_\phi}{r\sin\theta}\frac{\partial\Phi}{\partial\phi}$$

$$\nabla\cdot\mathbf{D} = \frac{1}{r^2}\frac{\partial}{\partial r}(r^2D_r) + \frac{1}{r\sin\theta}\frac{\partial}{\partial\theta}(\sin\theta D_\theta) + \frac{1}{r\sin\theta}\frac{\partial D_\phi}{\partial\phi}$$

$$\nabla\times\mathbf{H} = \frac{\mathbf{a}_r}{r\sin\theta}\left[\frac{\partial}{\partial\theta}(H_\phi\sin\theta) - \frac{\partial H_\theta}{\partial\phi}\right]$$

$$+ \frac{\mathbf{a}_\theta}{r}\left[\frac{1}{\sin\theta}\frac{\partial H_r}{\partial\phi} - \frac{\partial}{\partial r}(rH_\phi)\right] + \frac{\mathbf{a}_\phi}{r}\left[\frac{\partial}{\partial r}(rH_\theta) - \frac{\partial H_r}{\partial\theta}\right]$$

$$\nabla^2\Phi = \frac{1}{r^2}\frac{\partial}{\partial r}\left(r^2\frac{\partial\Phi}{\partial r}\right) + \frac{1}{r^2\sin\theta}\frac{\partial}{\partial\theta}\left(\sin\theta\frac{\partial\Phi}{\partial\theta}\right) + \frac{1}{r^2\sin^2\theta}\frac{\partial^2\Phi}{\partial\phi^2}$$

$$\nabla^2\mathbf{A} = \mathbf{a}_r\left[\nabla^2A_r - \frac{2}{r^2}\left(A_r + \cot\theta A_\theta + \csc\theta\frac{\partial A_\phi}{\partial\phi} + \frac{\partial A_\theta}{\partial\theta}\right)\right]$$

$$+ \mathbf{a}_\theta\left[\nabla^2A_\theta - \frac{1}{r^2}\left(\csc^2\theta A_\theta - 2\frac{\partial A_r}{\partial\theta} + 2\cot\theta\csc\theta\frac{\partial A_\phi}{\partial\phi}\right)\right]$$

$$+ \mathbf{a}_\phi\left[\nabla^2A_\phi - \frac{1}{r^2}\left(\csc^2\theta A_\phi - 2\csc\theta\frac{\partial A_r}{\partial\phi} - 2\cot\theta\csc\theta\frac{\partial A_\theta}{\partial\phi}\right)\right]$$

In the preceding three sets of vector operations in the various coordinates systems, the following definitions are valid:

$\nabla \phi \equiv$ gradient of a scalar $\phi$.

$\nabla \cdot \mathbf{D} \equiv$ divergence of a vector **D**.

$\nabla \times \mathbf{H} \equiv$ curl of a vector **H**.

$\nabla^2 \phi \equiv$ Laplacian of a scalar $\phi$.

$\nabla^2 \mathbf{A} \equiv$ Laplacian of a vector **A**.

## C.2 Vector Identities

$$\nabla(\Phi\psi) = \Phi\,\nabla\psi + \psi\,\nabla\Phi$$

$$\nabla \cdot (\psi\mathbf{A}) = \mathbf{A} \cdot \nabla\psi + \psi\nabla \cdot \mathbf{A}$$

$$\nabla \cdot (\mathbf{A} \times \mathbf{B}) = \mathbf{B} \cdot \nabla \times \mathbf{A} - \mathbf{A} \cdot \nabla \times \mathbf{B}$$

$$\nabla \times (\Phi\mathbf{A}) = \nabla\Phi \times \mathbf{A} + \Phi\nabla \times \mathbf{A}$$

$$\nabla \times (\mathbf{A} \times \mathbf{B}) = \mathbf{A}\nabla \cdot \mathbf{B} - \mathbf{B}\nabla \cdot \mathbf{A} + (\mathbf{B} \cdot \nabla)\mathbf{A} - (\mathbf{A} \cdot \nabla)\mathbf{B}$$

$$\nabla \cdot \nabla\Phi = \nabla^2\Phi$$

$$\nabla \cdot \nabla \times \mathbf{A} = 0$$

$$\nabla \times \nabla\Phi = 0$$

$$\nabla \times \nabla \times \mathbf{A} = \nabla(\nabla \cdot \mathbf{A}) - \nabla^2\mathbf{A}$$

$$\nabla(\mathbf{A} \cdot \mathbf{B}) = (\mathbf{A} \cdot \nabla)\mathbf{B} + (\mathbf{B} \cdot \nabla)\mathbf{A} + \mathbf{A} \times (\nabla \times \mathbf{B}) + \mathbf{B}(\nabla \times \mathbf{A})$$

$$\mathbf{A} \cdot \mathbf{B} \times \mathbf{C} = \mathbf{B} \cdot \mathbf{C} \times \mathbf{A} = \mathbf{C} \cdot \mathbf{A} \times \mathbf{B}$$

$$\mathbf{A} \times (\mathbf{B} \times \mathbf{C}) = \mathbf{B}(\mathbf{A} \cdot \mathbf{C}) - \mathbf{C}(\mathbf{A} \cdot \mathbf{B})$$

Stokes's Theorem $\int_S \nabla \times \mathbf{A} \cdot d\mathbf{S} = \oint_l \mathbf{A} \cdot d\mathbf{l}$ Open surface S bounded by line l

Divergence Theorem $\int_v \nabla \cdot \mathbf{A}\, dv = \oint_S \mathbf{A} \cdot d\mathbf{S}$ volume v bounded by surface s

# INDEX